The Man of Numbers

BY THE SAME AUTHOR

The Math Gene

The Numbers Behind Numb3rs

The Unfinished Game

The Man of Numbers

Fibonacci's Arithmetic Revolution

KEITH DEVLIN, PH.D.

Walker & Company
New York

Images on page 24 from George Gheverghese Joseph, *The Crest of the Peacock* (London: Penguin, 1991).

Published by Walker Publishing Company, Inc., New York
A Division of Bloomsbury Publishing

All papers used by Walker & Company are natural, recyclable products made from wood grown in well-managed forests. The manufacturing processes conform to the environmental regulations of the country of origin.

LIBRARY OF CONGRESS CATALOGING-IN-PUBLICATION DATA
HAS BEEN APPLIED FOR.

ISBN: 978-0-8027-7812-3

Visit Walker & Company's website at www.walkerbooks.com

First U.S. edition published in 2011
This paperback edition published in 2012

Paperback ISBN: 978-0-8027-7908-3

1 3 5 7 9 10 8 6 4 2

Designed by Rachel Reiss
Typeset by Westchester Book Group
Printed in the U.S.A. by Quad/Graphics, Fairfield, Pennsylvania

Contents

Acknowledgments

A number of people helped me with the writing of this book. The most significant contributions came from Professors Rafaella Franci of the University of Siena, Jeffrey Oaks of Indianapolis University, and Barnabas Hughes of California State University at Northridge, who each read the entire manuscript at various stages of development and provided valuable feedback. As a mathematician not well versed in the history of the subject, I found their input invaluable. I also benefited enormously from an e-mail exchange with, and later a visit to, Judith Sigler Fell, the widow of the mathematician who translated *Liber abbaci* from Latin to English.

My colleague Professor Franco Montagna of the University of Siena provided me with the initial introduction to Professor Franci, and a lengthy conversation with her and her colleague Professor Paolo Pagli early in the project helped orient my research. Likewise, a conversation with Professor Giulo Barozzi of the University of Bologna also provided me a valuable initial overview of the task that faced me in trying to piece together the story of Leonardo's life and work. Professor Greg Adams of the Mathematics Department of Bucknell University originally outlined the history of Laurence Sigler's English translation of *Liber abbaci* and put me in touch with his widow. Serena Ferrando of Stanford University provided an English translation of one particularly important Italian source (Franci's seminal article, described in chapter 8) for which I did not trust my own translation ability. Christine Holmes of the San Jose State University Library provided considerable assistance and support in conducting my researches in Italy.

Finally, I want to thank my literary agent, Ted Weinstein, who is always an enthusiastic and solid supporter of my efforts, and who consistently goes the extra mile to find the perfect publisher for my work, and George Gibson, my publisher at Walker & Company, who contributed significantly to the final form of the narrative.

CHAPTER 0

Your Days Are Numbered

TRY TO IMAGINE A DAY WITHOUT numbers. Never mind a day; try to imagine getting through the first hour without numbers: no alarm clock, no time, no date, no TV or radio, no stock market report or sports results in the newspapers, no bank account to check. It's not clear exactly where you are waking up either, for without numbers modern housing would not exist.

The fact is, our lives are totally dependent on numbers. You may not have "a head for figures," but you certainly have a head full of figures. Most of the things you do each day depend on and are conditioned by numbers. Some of them are obvious, like the ones listed above; others govern our lives behind the scenes. The degree to which our modern society depends on numbers that are hidden from us was made clear by the worldwide financial meltdown in 2008, when overconfident reliance on the advanced mathematics of futures predictions and the credit market led to a total collapse of the global financial system.

How did we—as a species and as a society—become so familiar with and totally reliant on these abstractions our ancestors invented just a few thousand years ago? As a mathematician, I had been puzzled by this question for many years, but for most of my career as a university professor of mathematics, the pressures of discovering new mathematics and teaching mathematics to new generations of students did not leave me enough time to look for the answer. As I grew older, however, and came to terms with the unavoidable fact that my abilities to do original mathematics were starting to wane a bit—a process that for most mathematicians starts around the age of forty (putting mathematics in the same category as many sporting activities)—I started to spend more time looking into the origins of the subject I have loved with such passion since I made the transition from "It's boring" to "It's unbelievably beautiful" around the age of sixteen.

For the most part, the story of numbers was easy to discover. By the latter part of the first millennium of the Current Era, the system we use today to write numbers and do arithmetic had been worked out—expressing any number using just the ten numerals 0, 1, 2, 3, 4, 5, 6, 7, 8, 9, and adding, subtracting, multiplying, and dividing them by the procedures we are all taught in elementary school. (Units column, tens column, hundreds column, carries, etc.) This familiar way to write numbers and do arithmetic is known today as the Hindu-Arabic system, a name that reflects its history.

Prior to the thirteenth century, however, the only Europeans who were aware of the system were, by and large, scholars, who used it solely to do mathematics. Traders recorded their numerical data using Roman numerals and performed calculations either by a fairly elaborate and widely used fingers procedure or with a mechanical abacus. That state of affairs started to change soon after 1202, the year a young Italian man, Leonardo of Pisa—the man whom a historian many centuries later would dub "Fibonacci"—completed the first general purpose arithmetic book in the West, *Liber abbaci*, that explained the "new" methods in terms understandable to ordinary people (tradesmen and businessmen as well as schoolchildren).[*] While other lineages can be traced, Leonardo's influence, through *Liber abbaci,* was by far the most significant and shaped the development of modern western Europe.

Leonardo learned about the Hindu-Arabic number system, and other mathematics developed by both Indian and Arabic[†] mathematicians, when his father brought his young son to join him in the North African port of Bugia (now Bejaïa, in Algeria) around 1185, having moved there from Pisa to act as a trade representative and customs official. Years later, Leonardo's book not only provided a bridge that allowed modern arithmetic to cross the Mediterranean, but also bridged the mathematical cultures of the Arabic and European worlds, by showing the West the algebraic way of thinking that forms the basis of modern science and engineering (though not our familiar symbolic notation for algebra, which came much later).

What Leonardo did was every bit as revolutionary as the personal computer pioneers who in the 1980s took computing from a small group of "computer types" and made computers available to, and usable by, anyone. As with those pioneers, most of the credit for inventing and developing the methods Leonardo

---

* Leonardo published a completely revised and extended second edition of *Liber abbaci* in 1228. No copies of the first edition have survived, but three almost complete copies of the 1228 version exist, dating back to the same era, and are kept in libraries in Rome, Florence, and Siena. See chapter 9.

† Different people use the word "Arab" to mean different things. In this book, I use it in the commonly accepted scholastic sense to mean the peoples whose primary business or cultural language was Arabic—just as we speak of "the Greeks" to refer to the peoples whose primary cultural language was Greek. Used in this sense, the term "Arab" includes peoples from many nationalities, mostly, but not all, Muslim. Similarly, when I write of "Muslim scholars" I am referring to scholars who lived and worked within the Muslim culture, regardless of their race, national origin, or religious beliefs or practices.

described in *Liber abbaci* goes to others, in particular Indian and Arabic scholars over many centuries. Leonardo's role was to "package" and "sell" the new methods to the world.

Not only did the appearance of Leonardo's book prepare the stage for the development of modern (symbolic) algebra and hence modern mathematics, it also marked the beginning of the modern financial system and the way of doing business that depends on sophisticated banking methods. For instance, Professor William N. Goetzmann of the Yale School of Management, an expert on economics and finance, credits Leonardo as the first to develop an early form of present-value analysis, a method for comparing the relative economic value of differing payment streams, taking into account the time value of money. Mathematically reducing all cash flow streams to a single point in time allows the investor to decide which is the best, and the modern version of the present-value criterion, developed by the economist Irving Fisher in 1930, is now used by virtually all large companies in the capital budgeting process.[1]

THE ONLY PIECE of the story of numbers that was missing was an account of Leonardo himself and, apart from a few scholarly articles, of the nature of his book. History has relegated him to an occasional footnote. Indeed, his name is known today primarily in connection with the Fibonacci numbers, a sequence of numbers that arises from the solution to the rabbit problem,[*] one of many whimsical challenges he put in *Liber abbaci* to break the tedium of the hundreds of practical problems that dominate the book.

Part of the reason Leonardo has been overlooked, whereas

* See chapter 9.

comparable figures like Copernicus, Galileo, and Kepler were not, may be that to most laypersons science seems to serve a greater purpose than mathematics.

Another reason why generations may have overlooked Leonardo is that the change in society brought about by the teaching of modern arithmetic was so pervasive and all-powerful that within a few generations people simply took it for granted. There was no longer any recognition of the magnitude of the revolution that took the subject from an obscure object of scholarly interest to an everyday mental tool. Compared with Copernicus's conclusions about the position of Earth in the solar system and Galileo's discovery of the pendulum as a basis for telling time, Leonardo's showing people how to multiply 193 by 27 simply lacks drama.

The comparative neglect of Leonardo has no doubt been caused by two other factors. Very little was recorded about his life, discouraging biographies. And Leonardo was more a salesperson of modern arithmetic rather than its inventor. The mathematical advances he described in *Liber abbaci* were developed by others, and others also wrote books describing those mathematical ideas. In the world of scientific biography, the inventor tends to get the glory. But inventions—an idea, a theory, a process, a technology—need to be made accessible to the world. The personal computer on which I write these words, with its familiar windows, mouse-controlled pointer, and the like, was invented by brilliant teams of researchers at the Stanford Research Institute and the Xerox Palo Alto Research Center in the 1970s, but it was put into everyone's hands by a few pioneering entrepreneurs. The computer revolution would undoubtedly have happened anyway, just as we would have figured out the motion of the planets had Kepler not lived, and gravity without Newton. But the likes of Apple Computer's Steve Jobs and Microsoft's Bill Gates will always be linked to the rise of the personal computer, and in this way Leonardo should be linked to the rise of modern arithmetic.

WHAT LEONARDO BROUGHT to the mathematics he learned in Bugia and elsewhere in his subsequent travels around North Africa were systematic organization of the material, comprehensive coverage of all the known methods, and great expository skill in presenting the material in a fashion that made it accessible (and attractive) to the commercial people for whom he clearly wrote *Liber abbaci.* He was, to be sure, a highly competent mathematician—in fact, one of the most distinguished mathematicians of medieval antiquity—but only in his writings subsequent to the first edition of *Liber abbaci* in 1202 did he clearly demonstrate his own mathematical capacity.

Following the appearance of *Liber abbaci*, the teaching of arithmetic became hugely popular thoughout Italy, with perhaps a thousand or more handwritten arithmetic texts being produced over the following three centuries. Moreover, the book's publication, and that of a number of his other works, brought Leonardo fame throughout Italy as well as an audience with the Holy Roman Emperor Frederick II. Since the Pisan's writings were still circulating in Florence throughout the fourteenth century, as were commentaries on his works, we know that his legacy lived on long after his death. But then Leonardo's name seemed to be suddenly forgotten. The reason was the invention of movable-type printing in the fifteenth century.

Given the Italian business world's quick adoption of the new arithmetic, not surpisingly the first mathematics text printed in Italy was a fifty-two-page textbook on commercial arithmetic: an untitled, anonymous work known today as the *Aritmetica di Treviso* (*Treviso Arithmetic*), after the small town near Venice where it was published on December 10, 1478. Soon afterward,

Piero Borghi brought out a longer and more extensive arithmetic text, printed in Venice in 1484, that became a true bestseller, with fifteen reprints, two in the 1400s and the last one in 1564. Filippo Calandri wrote another textbook, *Pitagora aritmetice introductor*, printed in Florence in 1491, and a manuscript written by Leonardo da Vinci's teacher Benedetto da Firenze in 1463, *Trattato d'abacho*, was printed soon afterward. These early printed arithmetic texts were soon followed by many others.

Though *Liber abbaci* was generally assumed to be the initial source for many, if not all, of the printed arithmetic texts that were published, only one of them included any reference to Leonardo.* Luca Pacioli, whose highly regarded, scholarly abbacus book *Summa de arithmetica, geometria, proportioni et proportionalità* (All that is known about arithmetic, geometry, proportions, and proportionality) was printed in Venice in 1494, listed Leonardo among his sources, and stated, "Since we follow for the most part Leonardo Pisano, I intend to clarify now that any enunciation mentioned without the name of the author is to be attributed to Leonardo."

The general absence of creditation was not unusual; citing sources was a practice that became common much later, and authors frequently lifted entire passages from other writers without any form of acknowledgment. But without that one reference by Pacioli, later historians might never have known of the great Pisan's pivotal role in the birth of the modern world. Yet Pacioli's remark was little more than a nod to history, for a reading of the entire text shows that the author drew not from *Liber abbaci* itself but from sources closer to his own time. There is no indication he had ever set eyes on a copy of *Liber abbaci*, let alone read it. His citation of Leonardo reflects the fact that, at the

* Several earlier handwritten manuscripts did refer to Leonardo.

time, the Pisan was considered the main authority, whose book was the original source of all the others.

Despite the great demand for mathematics textbooks, *Liber abbaci* itself remained in manuscript form for centuries, and therefore inaccessible to all but the most dedicated scholars.* Not only was it much more scholarly and difficult to understand than many other texts; it was very long. Over time it became forgotten, as people turned to shorter, simpler, and more derivative texts. That one mention in Pacioli's *Summa* was the only clue to Leonardo's pivotal role in the dramatic growth of arithmetic.† It lay there, unnoticed, until the late eighteenth century, when an Italian mathematician called Pietro Cossali (1748–1815) came across it when he studied *Summa* in the course of researching his book *Origine, transporto in Italia, primi progressi in essa dell-algebra* (Origins, transmission to Italy, and early progress of algebra there).[2] Intrigued by Pacioli's brief reference to "Leonardo Pisano", Cossali began to look for the Pisan's manuscripts, and in due course learned from them of Leonardo's important contribution.

In his book, published in two volumes in 1797 and 1799, which many say is the first truly professional mathematics his-

---

* It was not printed until 1857, when Baron Baldassarre Boncompagni, an Italian bibliophile and medieval mathematical historian, had the manuscript typeset and published it in Rome. Boncompagni's *Liber abbaci* comprised the first volume of a two-volume, printed collection of all of Leonardo's works he compiled under the title *Scritti di Leonardo Pisano.* The second volume, containing all of Leonardo's other works, appeared in 1862. A printed English-language translation of *Liber abbaci,* by the American mathematician Laurence Sigler, was published in 2002. Based on Boncompagni's edition, it runs to 672 pages and is the only translation of Leonardo's text into a modern language.

† Discounting references in some handwritten manuscripts that would not be uncovered until the late twentieth century when scholars began to investigate Leonardo's heritage.

tory book written in Italy, Cossali concluded that Leonardo's *Liber abbaci* was the principal conduit for the "transmission to Italy" of modern arithmetic and algebra, and that the new methods spread first from Leonardo's hometown of Pisa through Tuscany (in particular Florence), then to the rest of Italy (most notably Venice), and eventually throughout Europe.[3] As a result, Leonardo Pisano, famous in his lifetime then completely forgotten, became known—and famous—once again. But his legacy had come extremely close to being forever lost.

The lack of biographical details makes a straight chronicle of Leonardo's life impossible. Where and when exactly was he born? Where and when did he die? Did he marry and have children? What did he look like? (A drawing of Leonardo you can find in books and a statue of the man in Pisa are most likely artistic fictions, there being no evidence they are based on reality.) What else did he do besides mathematics? These questions all go unanswered. From a legal document, we know that his father was called Guilichmus, which translates as "William" (the variant Guilielmo is also common), and that he had a brother named Bonaccinghus. But if Leonardo's fame and recognition in Italy during his lifetime led to any written record, it has not survived to the present day.

Thus a book about Leonardo must focus on his great contribution and his intellectual legacy. Having recognized that numbers, and in particular powerful and efficient ways to compute with them, could change the world, he set about making that happen at a time when Europe was poised for major advances in science, technology, and commercial practice. Through *Liber abbaci* he showed that an abstract symbolism and a collection of seemingly obscure procedures for manipulating those symbols had huge practical applications.

The six-hundred-page book Leonardo wrote to explain those ideas is the bridge that connects him to the present day. We may not have a detailed historical record of Leonardo the man, but we have his words and ideas. Just as we can come to understand great novelists through their books or accomplished composers through their music—particularly if we understand the circumstances in which they created—so too we can come to understand Leonardo of Pisa. We know what life was like at the time he lived. We can form a picture of the world in which Leonardo grew up and the influences that shaped his ideas. (In that we are helped by the survival to this day, largely unchanged, of many of the streets and buildings of thirteenth-century Pisa.) And we know how numbers were used prior to the appearance of *Liber abbaci*, and how the book changed that usage forever.

CHAPTER 1

A Bridge of Numbers

Liber abbaci TRANSLATES AS "Book of calculation". The intuitive translation "Book of the abacus" is both incorrect and nonsensical, inasmuch as Leonardo's book showed how to do arithmetic without the need for any such device as an abacus.* The distinction is reflected in Leonardo's spelling. The Latin and Italian word *abbacus* was used in medieval Italy from the thirteenth century onward to refer to the method of calculating with the Hindu-Arabic number system. The first known written use of the word *abbacus* with this spelling and meaning was in fact in the prologue of Leonardo's book. Thereafter, the word *abbaco* was widely used to describe the practice of calculating. A *maestro d'abbaco* was a person who was proficient

* The familiar calculating device comprising beads strung along wires attached to a wooden frame, although popularly called an "abacus", was not used in medieval Europe, but rather came from China, where it was called a *xuanpan*. That device is, therefore, sometimes more accurately described as a "Chinese abacus." The European "abacus" was a board or table (*tavola*) ruled with a series of parallel lines on which the user slid counters (or "jetons") to represent numbers. It was used extensively all over Europe in the Middle Ages and could still be found in use in some places as late as the eighteenth century.

in arithmetic. In fact, *abbaco* still has that as its primary (preferred) meaning in contemporary Italian.*

Medieval authors did not usually give their works titles. The name we use today for Leonardo's book comes from his opening statement:

Here begins the Book of Calculation
Composed by Leonardo Pisano, Family Bonacci,
In the Year 1202

In later writings, he also referred to the work as *Liber numerorum* and, in the dedicatory letter for his book *Flos*, as his *Liber maior de numero*. In chapter 5 of another of his books, *De practica geometrie*, written between the publication of the two editions of *Liber abbaci*, he used the title *Liber abbaci* again. "Since at the beginning of the treatise I had promised to discuss how to find cube roots, a topic to which I gave special attention in *Liber abbaci*, I rewrote the material for a regular chapter here."[1] In addition to its appearance in the opening statement, the word *abbaci* (the Latin genitive of *abbacus*) occurs in *Liber abbaci* three other times: in the prologue, where Leonardo described how he pursued *studio abbaci* "for some days" in Bugia; in chapter 12, when he stated he would treat a *questionibus abbaci*; and toward the end of the book, when he explained that his numerical determination of the approximate square root of 743 was done *secundum abbaci materiam*.

In addition to confusion over the book's title, there is uncertainty as to the full and correct name of the author. According to the tradition of the time, he would have been known as

* Historians have not been consistent over this spelling distinction, and it is unfortunate that the English translation of *Liber abbaci* uses the spelling with one "b".

"Leonardo Pisano" (Leonardo of Pisa). In his opening statement, he referred to himself as *filius Bonacci*, a Latin phrase that translates literally as "son of Bonacci". But Bonacci was not his father's name, so we should perhaps translate the phrase as "of the Bonacci family".[2] In any event, the Latin phrase *filius Bonacci* is the origin of Leonardo's present-day nickname "Fibonacci", coined by the historian Guillaume Libri in 1838. A further name Leonardo occasionally used to refer to himself was "Bigollo", a Tuscan dialect term sometimes used to refer to a traveler, but that meaning may be a coincidence. (In some old dialects the word also meant "blockhead", but since Leonardo used the term himself, that surely was not his intended meaning.)

Leonardo first encountered the number system that would fascinate him when, as a youth possibly no more than fifteen years of age, he left his childhood home in Pisa to join his father in the southern Mediterranean city of Bugia. There, in Muslim North Africa, he came into contact with Arabic-speaking traders and scholars who revealed to him a remarkable system for writing numbers and performing calculations. They were not its discoverers, for the system had its origins much earlier in India. By using it in their trading, Arab merchants then transported it northward along the Silk Road to the shores of the Mediterranean, together with other, more tangible products of the Orient, such as silk, spices, ointments, and dyes.

Humans had been counting for many thousands of years before the first number system was developed. Early counting, which goes back at least thirty-five thousand years, was done by scratching tally marks on a stick or bone. The oldest known example is the Lebombo bone, discovered in the Lebombo Mountains of Swaziland and dated to approximately 35,000 BCE, which consists of twenty-nine distinct notches deliberately cut into a baboon's fibula. It has been suggested that women used such notched bones to keep track of their menstrual cycles, making twenty-eight to thirty scratches on bone or stone, followed by a distinctive marker. Other examples of notched bones discovered in Africa and France, dated between 35,000 and 20,000 BCE, may have been early attempts to quantify time. The Ishango bone, found near the headwaters of the Nile in northeastern Congo and perhaps twenty thousand years old, consists of a series of tally marks carved in three columns running the length of the bone. A common interpretation is that the Ishango bone was a six-month lunar calendar.

With tally marks you simply make a vertical mark to record each item in a collection:

| || ||| |||| ||||| |||||| et cetera.

Tally marks become hard to read once you have more than four or five items to count. A common way to reduce the complexity is to group the tally marks in fives, often by drawing a diagonal stripe across each group of five tallies. The Roman numeral system, used throughout the Roman Empire and still found today in certain specialized circumstances, was a more sophisticated version of this simple idea, involving a few additional symbols: V for five, X for ten, L for fifty, C for a hundred, and M for a thousand. For example, using this system, the number one thousand two hundred and seventy eight (1,278) can be written as MCCLXXVIII:

$$\begin{aligned} \text{MCCLXXVIII} &= \text{M} + \text{C} + \text{C} + \text{L} + \text{X} + \text{X} + \text{V} + \text{I} + \text{I} + \text{I} \\ &= 1000 + 100 + 100 + 50 + 10 + 10 + 5 \\ &\quad + 1 + 1 + 1 \\ &= 1278 \end{aligned}$$

Addition in the Roman system is fairly easy, since you simply group all like symbols. For example, to add MCCXXIII (1,223) to MCXII (1,112) you simply collect all the M's, all the C's, and all the I's, like this:

M	CC	XX	III
M	C	X	II
MM	CCC	XXX	IIIII

Occasionally, you might have to convert one group of symbols to a higher symbol, for example the five I's could be replaced by V, to write the answer as MMCCCXXXV (2,335). Subtraction too is relatively easy. But the only tolerable way to do multiplication is by repeated addition and division by repeated subtraction. For example, V times MMCIII can be computed by adding MMCIII to itself four times. This method only works in practice when one of the two numbers being multiplied is small, of course.

The impracticality of the Roman system for doing multiplication or division meant it was inadequate for many important applications that arose in commerce and trade, such as currency conversion or determining a commission fee for a transaction. And there is no way Roman numerals could form the basis for any scientific or technical work. Societies that wrote numbers in Roman numerals used elaborate systems of finger arithmetic or mechanical devices—various kinds of abacus—to perform the actual calculations, using the numerals simply to record the answers. Although systems of finger arithmetic could accommodate arithmetic calculations involving numbers up to 10,000, and some individuals became so expert in using an abacus that they could carry out a computation almost as fast as a person today using a calculator, this required considerable physical dexterity and expertise. Since there was no record of the calculation, the answer had to be taken on trust.

The number system we use today—the Hindu-Arabic system—was developed in India and seems to have been completed by around 700 CE. Indian mathematicians made advances in what would today be described as arithmetic, algebra, and geometry, much of their work being motivated by an interest in astronomy. The system is based on three key ideas: notations for the numerals, place value, and zero. The choice of ten basic number symbols—that is, the Hindus' choice of the base 10 for counting and doing arithmetic—is presumably a direct consequence of using fingers to count. When we reach ten on our fingers we have to find some way of starting again, while retaining the calculation already made. The role played by finger counting in the development of early number systems would explain why we use the word "digit" for the basic numerals, deriving from the Latin word *digitus* for finger.*

An oft-repeated, though unproven, explanation for the choice of symbols used to represent the numerals is that if you write them using straight lines—a reasonable restriction in the days when writing was done on clay tablets using a stylus—then the number of angles in each figure is the number the figure represents. This, of course, depends on exactly how you write each numeral. Here is one way that makes everything work out correctly:

0123456789

---

* For doing arithmetic, any other base of roughly the same size would work as well. The satirist, songwriter, and mathematician Tom Lehrer once quipped that base 8 arithmetic is no harder than base 10 "if you're missing two fingers."

The introduction of zero was a crucial step in the development of Hindu arithmetic and came after the other numerals. The major advantage of the Hindus' number system is that it is positional—the position of each numeral matters. This allows for addition, subtraction, multiplication, and even division using fairly straightforward and easily learned rules for manipulating symbols. But for an efficient place-value number system, you need to be able to show when a particular position has no entry. For example, without a zero symbol, the expression

1 3

could mean thirteen, or a hundred and three (103), or a hundred and thirty (130), or maybe a thousand and thirty (1,030). One could put spaces between the numerals to show that a particular column has no entry, but unless one is writing on a surface marked off into columns, one can never be sure whether a particular space denotes a zero entry or is just the space separating the symbols. Everything becomes much clearer when there is a special symbol to mark a space with no value.

The concept of zero took a long time to develop. Since the number symbols were viewed as numbers themselves—things you used to count the number of objects in a collection—0 would be the number of objects in a collection having no members, which makes no sense. Other societies were never able to make the zero breakthrough. For instance, long before the Indians developed their system, the Babylonians had a positional number system, based on 60. Vestiges of their system remain when we measure time and angles: 60 seconds equal one minute, 60 minutes one hour, 60 angular seconds equal one degree, and 360 ($= 6 \times 60$) degrees make a full circle. But the Babylonians did not have a symbol denoting zero, a limitation to their system they were never able to overcome.

The Hindus got to zero in two stages. First they overcame the problem of denoting empty spaces in place-value notation by drawing a circle around the space where there was a "missing" entry. This much the Babylonians had done. The circle gave rise to the present-day symbol 0 for zero. The second step was to regard that extra symbol just like the other nine. This meant developing the rules for doing arithmetic using this additional symbol along with all the others. This second step—changing the underlying conception so that the rules of arithmetic operated not on the numbers themselves (which excluded 0) but on symbols for numbers (which included 0)—was the key. Over time it led to a change in the conception of numbers to a more abstract one that includes 0. The zero breakthrough was made by a brilliant mathematician called Brahmagupta.

Born in 598 in northwest India, Brahmagupta lived most of his life in Bhillamala (modern Bhinmal in Rajasthan). In 628, when he was thirty years old, he wrote a mammoth (twenty-five chapters) treatise called *Brahmasphutasiddhanta* (The opening of the universe). He went on to become the head of the astronomical observatory at Ujjain, the foremost mathematical center of ancient India at the time, and in 665, at sixty-seven years old, he wrote another book on mathematics and astronomy, *Khandakhadyaka*.

Brahmagupta introduced the number zero in *Brahmasphutasiddhanta*, describing it as the answer you get when you subtract a number from itself. He worked out some basic properties that zero must have, such as:

> When zero is added to a number or subtracted from a number, the number remains unchanged; and a number multiplied by zero becomes zero.

He gave arithmetic rules for handling positive and negative numbers (including rules for zero) in terms of fortunes (positive numbers) and debts (negative numbers):

> A debt minus zero is a debt.
> A fortune minus zero is a fortune.
> Zero minus zero is a zero.
> A debt subtracted from zero is a fortune.
> A fortune subtracted from zero is a debt.
> The product of zero multiplied by a debt or fortune is zero.

NUMBERS ARE SO ubiquitous in modern life, so much a part of the structure of our daily world, that we take them for granted, failing to see how remarkable is the Hindu-Arabic system just for writing numbers, let alone calculating with them. We see the expression 13,049, for example, and we recognize at once that it is the number thirteen thousand and forty-nine. This understanding is notable on several levels. For one thing, it is much easier to read the symbolic expression (and know what number it means) than to read the description in words. Somehow, we feel that the symbolic version is the number, whereas the expression in words is just a description of the number. This is more than just our perception. In recent years, experimental psychologists have used laboratory techniques, together with studies of individuals with brain lesions that destroy number and language capacities, to demonstrate that our brains store numbers along with—and arguably through—the symbols that represent them.[3] Our sense of numbers depends on the symbols, and we cannot divorce the symbols from the numbers they represent.

Another remarkable thing about our number system is that

using just the ten symbols (or digits) 0, 1, 2, 3, 4, 5, 6, 7, 8, 9, we can represent any of the infinitely many positive whole numbers. That efficiency is achieved by making use of the position each digit occupies. The rightmost digit in any number expression represents itself. The next one to the left represents that many tens, the next along that many hundreds, et cetera. Thus, in the number expression 538 the 8 denotes eight, the 3 denotes three tens, or thirty, and the 5 denotes five hundreds, or five-hundred. Reading left-to-right, the expression as a whole denotes the number *five hundred* and *thirty* and *eight*, that is, *five hundred and thirty-eight*. Symbolically:

$$538 = (5 \times 100) + (3 \times 10) + (8)$$

The zero symbol, 0, allows us to skip a column. For instance, 207 denotes two hundreds plus *no* tens plus seven units, that is, two hundred and seven.

On the face of it, this discussion may seem circular, like saying that the word "blue" means the color blue, but this just confirms how familiar to us are numbers and the way we write them. Numbers—things we use to count collections of objects—are not at all the same as the symbols we use to represent them. For example, there is only one *number* three, but many ways to represent it: "3" (in symbols), "three" (in English), "tres" (in French), "drei" (in German), "tre" (in Spanish), et cetera.

The modern symbolic notation for numbers and arithmetic is the world's only truly universal language. Writing numbers as we do makes arithmetic—adding, subtracting, multiplying, and dividing pairs of numbers—routine. Provided you start out by placing the two numbers correctly—in the case of addition, subtraction, or multiplication, one number beneath the other, with their digits lined up vertically starting from the right—the rest of the calculation is routine and mechanical.

This remarkable number system gradually spread northward from India as the traders who traveled between North Africa and the Orient learned of it and started to use it for their commercial transactions. By the end of the twelfth century, the Hindu-Arabic system was in use in trading ports all along the southern shore of the Mediterranean. Then Leonardo took it across the water to Italy.

In fact, the Hindu-Arabic numerals and Hindu-Arabic arithmetic had appeared in Italy—separately—before Leonardo was born, but both had languished, the former regarded as little more than a curiosity, the latter unknown outside a small circle of scholars.

After the Arabs invaded Spain in 711 CE, there was regular traveling and trade between Spain and the Arabic world, together with the exchange of books and information. The oldest dated Latin manuscript containing the Hindu-Arabic numerals—though it did not show how to use them in calculations—is the *Codex Vigilanus,* a compilation of historical documents written in 976 in Spain and found in the monastery of Albelda in the Rioja (Asturias). The manuscript is a copy made by the monk Vigila of an earlier work, the *Etymologies* of Isidore of Seville. As was common practice, Vigila incorporated his own commentaries, and he preceded a description of the Hindu-Arabic numerals with these words: "We must know that the Indians have a most subtle talent and all other races yield to them in arithmetic and geometry and the other liberal arts. And this is clear in the 9 figures with which they are able to designate each and every degree of each order (of numbers). And these are the forms."

Vigila may have learned of the numerals from Christians educated in al-Andalus (Mozarabs) who had emigrated to northern

Spain. Or perhaps he saw them used on the variant of the abacus board developed by the Frenchman Gerbert d'Aurillac, who in later life became Pope Sylvester II. In 967, when Gerbert was roughly the same age Fibonacci was when he went to Bugia, the young Frenchman traveled to Catalonia, where he studied mathematics for three years under the supervision of Hatto, bishop of Vich. While in Spain, Gerbert learned about the Hindu-Arabic numerals and used them in an attempt to improve the efficiency of the abacus board.

Gerbert's monastic abacus had twenty-seven columns (three for fractions). His main innovation was to use single counters marked with symbols in place of groups of counters—one symbol to denote that the marked counter stood on its own, another symbol to show that it stood for two original counters, another symbol to indicate three originals, and so on up to a symbol showing that the marked counter stood for nine original counters. These marked counters were called *apices*, from the Latin *apex*, presumably because the counters were cone-shaped and Gerbert's markings adorned the apex. He had a thousand such apices carved from horn. The symbols that adorned the apices were an early form of the Hindu-Arabic numerals—absent a zero, since an abacus board represented a zero by a column with no counter in it, and thus did not require a symbol.

Gerbert's abacus board remained popular for teaching arithmetic until at least the mid-twelfth century. It was, however, not used by merchants; for although it demonstrated place value, with a single symbol in each column, it was not particularly efficient, since calculations required constant swapping of symbols. In adopting the Hindu-Arabic numerals as mere markers, Gerbert missed the real power of Hindu-Arabic arithmetic. The appearance of Gerbert's board does, however, seem to have been the first time the Christian Western world saw the Hindu-Arabic numerals. As one of Gerbert's pupils later wrote of his teacher, "He used nine symbols, with which he was able to express every number."[4] Two manuscripts depicting Gerbert's abacus mention that he "[gave] to the Latin world the numbers of the abacus and their shapes."[5] A century and a half later, William of Malmesbury declared that Gerbert had "snatched the abacus from the Arabs."[6]

GERBERT WAS NOT alone in failing to recognize the power that lay behind the new symbols, and for more than a hundred years Europeans viewed them as little more than curious marks on counters. Though the *Codex Vigilanus* was written in Spain, that country did not adopt the new way of doing arithmetic until long after Italy. The earliest French manuscript that even described the numerals, let alone made use of them, was not written until 1275, long after the appearance in Italy of *Liber abbaci*, and the new arithmetic did not start to make headway among French merchants until many decades later.

The Hindu-Arabic numerals—purely as symbols for denoting numbers—had reached Pisa by 1149, when they were used to make the entries in the "Tables of Pisa," astronomical tables believed to be Latin translations of some Arabic tables written in the late tenth century.[7] It seems unlikely that Leonardo came across the new numerals while growing up in Pisa, however. When the new symbols first came into Europe, they were written the way the Arabs did—the so-called Eastern form—but when Leonardo wrote *Liber abbaci* he formed them in a different fashion known as the "Western form", which is the one we are familiar with today.

The methods of Hindu-Arabic arithmetic had also reached Europe before Leonardo's lifetime, but no one saw their practical significance. Half a century before Leonardo journeyed to

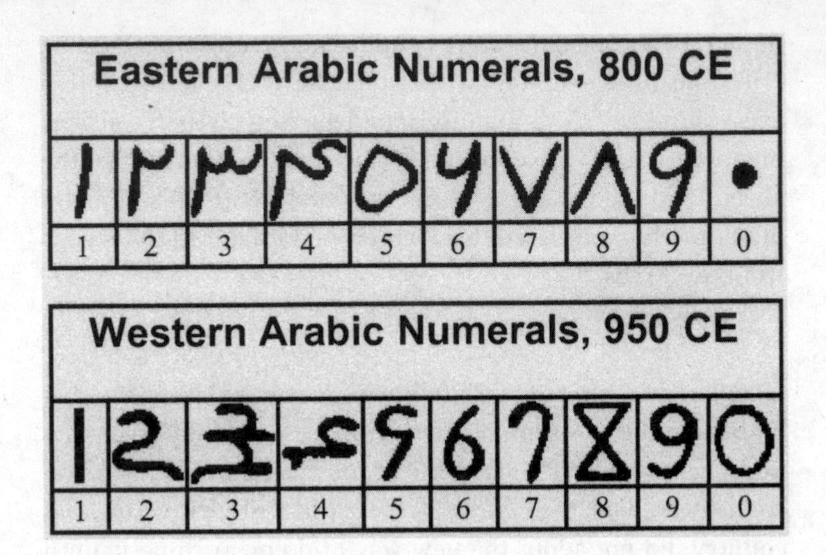

North Africa, European scholars had translated into Latin two important Arabic manuscripts, written by the ninth-century Persian mathematician Abū 'Abdallāh Muḥammad ibn Mūsā al-Khwārizmī (ca. 780–ca. 850 CE).

The first, written around 825, described Hindu-Arabic arithmetic. Its original title is not known, and it may not have had one.[8] No original Arabic manuscripts exist, and the work survives only through a Latin translation, which was most likely made in the twelfth century by Adelard of Bath. The original Latin translation did not have a title, either, but it was given one when it was printed in the nineteenth century: *Algoritmi de numero Indorum* (al-Khwārizmī on the Hindu art of reckoning).[9] The Latinized version of al-Khwārizmī's name in this title (*Algoritmi*) gave rise to our modern word "algorithm" for a set of rules specifying a calculation. The work is also referred to occasionally by the first two words with which it starts: *Dixit algorizmi* (So said al-Khwārizmī), and still another title is *On the Calculation with Hindu Numerals*, but it is most often referred to simply as "al-Khwārizmī's *Arithmetic*."

Al-Khwārizmī's second book, completed around 830, was *al-Kitab al-mukhtasar fi hisab al-jabr wa'l-muqabala*, which translates literally as "The abridged book on calculation by restoration and confrontation", or more colloquially "The abridged book on algebra".[10] It is an early treatise on what we now call "algebra", that name coming from the term *al-jabr* in the title. The phrase *al-jabr wa'l-muqabala* translates literally as "restoration and confrontation", or more loosely as "balancing an equation". Scholars today usually refer to this book as "al-Khwārizmī's *Algebra*". In *Algebra*, al-Khwārizmī developed a systematic approach to solving linear and quadratic equations, providing a comprehensive account of solving polynomial equations up to the second degree.

Whereas al-Khwārizmī wrote his books for merchants and businessmen, the European Latin translations were primarily written for, and largely read only by, other scholars. Interested solely in the benefits of the system within mathematics, the translators did not see any significance for the world of commerce. That important observation had to wait until the young Leonardo Pisano traveled to North Africa.

CHAPTER 2

A Child of Pisa

THOUGH LEONARDO WAS BORN into a wealthy family with influential friends, and was well known for *Liber abbaci* and several other books when he died, there are few historical records relating to him. We know he was born sometime around 1170 CE, but we do not know the exact year, and we are not completely sure where. Most likely, it was in Pisa, but in any event that was where he spent most of his childhood. According to the custom for naming at that time, he would have been known publicly as Leonardo Pisano. He was a contemporary of Bonanno Pisano (Bonanno of Pisa), the engineer who started the construction of the Leaning Tower. Guilichmus, or Guilielmo (William), Leonardo's father, was a Pisan merchant turned customs official, which meant that the young Leonardo grew up in the company of the sons and daughters of other merchants—a childhood influence that was to have far-reaching consequences.

To be born in Pisa in the twelfth century was to enter the hub of the Western world. And to grow up in a Pisan merchant family was to be a member of what was then the most important sector of society. When Leonardo was born, Italy was a center of the vastly important, and still rapidly growing, international trade between the countries that fanned out from the Mediterranean Sea. Pisa, along with Italy's other maritime cities, Genoa to the north and Venice on the northeastern coast, dominated the trade, and their ships sailed constantly from one Mediterranean port to another. The merchants in those three cities were the key figures in shaping the development of a new, more cosmopolitan world.[1]

Evidence of Pisa's origins stretches back almost two thousand years BCE, when it served as a transit port for Greek and Phoenician trade to and from Gaul. Later, the Romans also used it as a port. But not until another thousand years had passed did Pisa begin to rise to the prominence it enjoyed when Leonardo was born. Travelers today who approach Pisa by train from Florence notice that as they near their destination, the beautiful rolling hills of the Chianti wine region give way to a large flat plain, which stretches beyond Pisa to the sea. After heavy rains, the land here floods regularly, a lasting reminder of why Pisa had become a port in the first place: This modern-day floodplain is where, in Roman times and earlier, Pisa's harbor used to be. In pre-Christian times, the Arno River, which today divides the city, opened up to a large lagoon just to the east, providing a natural port. The Romans called it the "Sinus Pisanus", although they were not the first to berth ships here.

By Leonardo's time, however, the lagoon had silted up, and Pisa's status as a major shipping port alongside Genoa and Venice was sustained purely by the expertise and connections of its citizens, not its location. Indeed, sometimes in dry weather the Arno River became too shallow for larger ships to reach the city. Broad-beamed sailing ships and seagoing barges could generally get through, but the bigger vessels had to berth at Porto Pisano ("Port of Pisa", nowadays part of the busy Mediterranean port-city of Livorno), several miles to the south of the Arno's mouth, along the seacoast. Their cargoes were then unloaded and carried into Pisa on narrow, oared galleys or on river flatboats propelled by hand using a pole.

Other changes were also affecting the lives of the Pisans in Leonardo's time. During the tenth century, as the five hundred years of cultural stagnation known as the Dark Ages came to an end, European society began to develop and prosper once again. New farming techniques were introduced, populations started to expand, and national and international commerce began to develop. With few roads available, and most of those of poor quality, trading was carried out largely by river and sea transport. As a result, the bulk of Western civilization was clustered around the shores of the Mediterranean.

From the tenth century onward, Pisa started to spread beyond its ancient Roman walls, with towers rising to the east and west, and to the south across the Arno. By the second half of the twelfth century, when Leonardo was growing up, a new, heavily fortified city wall was being constructed, to protect the city from attack both by Muslims—this was the era of the Crusades—and from rival Italian cities, which often attacked one another as part of an ongoing political struggle between the Holy Roman Emperor Frederick II and the pope.

Visitors who stroll around today's Pisa will occasionally come across buildings dating back to Leonardo's time: rectangular towers, built of stone or brick, rising three or more stories high. With constant feuding between rival families, a tower provided any Pisan family of means with a refuge as much as a home. The ground floor was often a shop or a storeroom for oil, wine, tools, and supplies; the second floor served as the main living area, and perhaps a bedroom. The kitchen was usually on the top floor, to allow smoke to escape easily. The common Pisan boast that the city had ten thousand such towers was surely a huge exaggeration, but as a child in a wealthy merchant family, Leonardo almost certainly grew up in such a building.

The well-known Italian surname Visconti has its origins in Pisan history of those times. In its early years, Pisa was officially part of Tuscany, which was ruled by a marquis who owed his allegiance to the emperor. The marquis's representative in Pisa was called a vice-count, or viscount. Over time, the viscounts began to keep the position within their own family, eventually taking the name of the office as their family name: the Visconti family. During Leonardo's childhood, the Viscontis' towers dominated the central quarter of the city—the Mezzo—although other families would later grow powerful enough to challenge their position.

Leonardo grew up during a period of enormous change. In Bologna, seventy miles northeast of Pisa, the first "university" was established in 1088. In Salerno, in the south, the first medical school was formed, attracting students from many different countries. Scholars in Pisa, Florence, and Siena were busy translating into Latin the great works of the Greeks: Euclid, Apollonius, Archimedes, Aristotle, and Galen. Of particular note, Ptolemy's astronomy treatise *Almagest,* one of the most comprehensive Greek works, was translated in Palermo in 1160 and again in Toledo in 1175. And communication between the different cities was made more efficient by the introduction of a postal service, one of the first in Europe.

Late in the eleventh century, scholars had discovered in a library in Pisa a complete and intact manuscript of the *Corpus iuris civilis,* the "Body of civil law" compiled by the emperor Justinian in the sixth century. In addition to becoming the focus of much academic study throughout the following century, by the time Leonardo was growing up, the rules and principles laid out in the treatise had already found their way into the Italian system of government.*

New financial institutions—banks—emerged during the twelfth century, growing in a few short decades from individual entrepreneurs who traveled around the country to the markets and trading fairs, carrying sacks of silver coins, to organized, and invariably wealthy, limited-liability collectives with fixed premises. In the early days, the roaming financiers had laid out their coins on wooden benches or banks—the Latin term was *banca*—so people started to call them "bankers". By Leonardo's day, the banks offered loans and issued letters of credit.† Groups of businesspeople and merchants would join forces and pool their resources to form limited-liability companies. The leaders would often hold their important meetings seated around a large dining table, or board, giving rise to the modern term "board of directors".

Trading was brisk between the European nations on the northern shores of the Mediterranean and the Arabic countries to the south. European merchants sold wool, cloth, timber, iron, and other metals to the Arabs. In the opposite direction, spices, medicines, ointments, cosmetics, dyes, tanning agents, and other goods were shipped across the Mediterranean into Europe. Many of these items originated in India and Ceylon; their long journey took them northwest to the head of the Persian

* Today, in large part as a result of that one document discovered in Pisa, Roman law provides the foundation of the civil law systems of many countries, including all the countries of continental Europe.

† The most prominent bankers were those from the northern Italian region of Lombardy, particularly those of its capital, Milan, which is why the major financial streets in both London and San Francisco are called Lombard Street.

Gulf, then by boat up the Tigris to Baghdad or Mosul, and by camel on to Syria or to the Red Sea ports of Egypt and to the Nile.

Dominating the Mediterranean trade were the ships from Pisa, Genoa, and Venice. While most of Italy was at that time under the rule of either the Holy Roman Emperor, the king of Sicily, or the pope, these three great seafaring cities functioned in many respects like nation-states, as did the inland cities of Florence and Milan. With strong armies and navies, not only were the Italian city-states able to fend off attacks from land and sea; they also gained strongholds elsewhere, including some key ports on the shores of North Africa. By the middle of the twelfth century, Pisa, whose population then numbered about ten thousand, had colonies, port privileges, or consular representatives all around the edge of the Mediterranean. Pisan merchants traded with the vast Muslim community that stretched in a crescent from Persia (present-day Iran), around the eastern and southern shores of the Mediterranean, and on as far as southern Spain.

Because of the wealth the traders brought to Pisa, Leonardo also grew up during a period of great cultural development. In many major Italian cities, masons, sculptors, and architects were constructing great architectural monuments. In Pisa, the most ambitious project was being undertaken in the northwest corner of the city, in what was to become the Piazza dei Miracoli—the Square of Miracles. There, a complex of buildings belonging to the diocese had been under construction for more than a century. When Leonardo was born, the cathedral and the baptistery were complete, although the baptistery dome would not be added for another century. But what was to prove the most interesting construction was just getting under way: the bell tower.

Marble blocks for the tower were brought in by barge from quarries in the mountains. Heavy carts then transported them to the building site, where they were given a final dressing by stonecutters before being hoisted into position by cranes and mortared together. Just when the tower reached three of its intended eight stories, the foundations shifted and the tower started to topple to one side. This was not unusual in cities such as Pisa that were built on soft ground. Without the benefits of present-day soil science, buildings often leaned and toppled. Bonanno Pisano, the engineer in charge of building the bell tower, struggled to avoid a complete collapse. In order to bring the tower nearer to vertical, he made the upper stories slightly taller on one side to compensate for the lean. But the additional weight of the extra masonry on that side simply caused the foundations to sink even further. When the tower was eventually completed in the fourteenth century, it still leaned—as it does today.

Leonardo's childhood was spent surrounded by numbers in action. This would have been particularly apparent along the banks of the Arno River, which runs from east to west, dividing Pisa in two.* At each end of the city was a customhouse. The one at the western wall, being nearest to the sea, handled vessels arriving from abroad. A typical incoming cargo might consist of sacks of grain from other parts of Italy, salt from Sardinia, bales of squirrel skins from Sicily, goatskins from North Africa, or ermine from Hungary. Some vessels had large doors in their sterns, which could be opened to allow horses from Provence to be led ashore. Particularly valuable imports were alum, destined for Pisa's leather industry, dyes for the textile manufacturers of Italy and northwest Europe, and spices from the Far East. Goods destined for Florence were transferred from ship to barge for the voyage farther up the Arno. When all the cargo

* Today, each bank carries a major road, bustling with traffic—one of those roads being called the Lung'arno Leonardo Fibonacci for part of its length.

had been taken ashore, Pisan workers would reload the ships with goods for export: barrels of Tuscan wine and oil, bales of hemp and flax, and bars of smelted iron and silver.

The eastern customhouse, facing inland, served traffic from upriver. Shallow-draft boats and barges brought farm produce in from the countryside or goods from Florence and other inland towns to sell in Pisa's year-round market. Just beyond the customhouse was the Long Ford, where the Arno broadened and grew shallow enough to ride a horse across in low water.

Next to the western customhouse was the shipyard. Shipbuilding was a booming industry in twelfth-century Pisa, and its skilled craftsmen built ships not just for Italian clients but for French and North African customers as well. Specially felled timbers were brought in by barge from the wooded uplands, unloaded by giant cranes, and cut into planks in a large pit using a pit saw. Two men operated the saw, one standing on the ground above, the other below in the pit. They pushed and pulled the huge vertical blade, slicing through the log, as others shoved it lengthwise against the saw. The timbers were shaped using heavy hammerlike adzes with curved iron blades. Despite the crude nature of their tools, skilled shipwrights were able to fashion the ships' timbers with remarkable precision, so as to avoid having to overlap the planks, as was common with most other shipbuilders at that time. To make the ship seaworthy, caulkers worked their way over the entire hull, sealing holes and cracks with hot pitch.

At the water's edge by the Piazza San Nicola and across the river in the Kinsica quarter, tanners took raw hides shipped in from North Africa, and scraped them over a section of a tree trunk to remove hair and flesh. Then they soaked them in cold water and myrtle—the source of tanning's distinctive smell—rubbing and beating them every day for up to six months, gradually transforming the raw skins into fine leather, ready to be cut and sewn into hats, belts, trousers, and other garments. Another commodity brought to Pisa by barge in Leonardo's time was wool, which was just starting to replace leather for clothing. Spinning, weaving, fulling (treating the woven cloth for softness and resilience), and dyeing were traditionally country industries, as was the sale of woolen cloth, but during the early thirteenth century these industries began to shift to the city.

Scattered along the riverbanks were dozens of colorful tents and improvised huts, temporary places of business erected by foreign merchants—Turks, Arabs, Libyans, and others—to display silks, carpets, vases, and other wares for sale.

Underlying all this activity—in the customhouses, on the wharves, in every place of business—were numbers. Merchants measured out their wares and negotiated prices; customs officers calculated taxes to be levied on imports; scribes and stewards prepared ships' manifests, recording the values in long columns using Roman numerals. They would have put their writing implements to one side and used either their fingers or a physical abacus to perform the additions, then picked up pen and parchment once again to enter the subtotals from each page on a final page at the end. With no record of the computation itself, if anyone questioned the answer, the entire process would have to be repeated.

As we look back with the hindsight of history, it is tempting to conjecture that the commercial activities young Leonardo observed along the banks of the Arno enabled him to later recognize in the powerful arithmetic methods he observed being used in Bugia the potential to revolutionize world trade. In any event, when Leonardo wrote *Liber abbaci*, he clearly did so primarily for the merchants, based on the contents and structure of the book. He took pains to explain the concepts in a way that those highly practical men could understand, presenting many examples from everyday commercial life.

CHAPTER 3

A Mathematical Journey

FOR SOMEONE WHOSE INTELLECTUAL work was to change the course of history, Leonardo's schooling would have been decidedly basic—for the simple reason that there was nothing else available. In the late twelfth century, education throughout Europe was in the hands of the monasteries and the cathedrals. The curriculum, if it can be so called, comprised little more than learning to read and write, and to write numbers using the Roman system. Leonardo may also have been taught some practical geometry.[1]

He would have attended school between ten and twelve years of age. The school was most likely in the cathedral in Pisa. It would have had no desks or chairs; the pupils—all boys—would have sat cross-legged on the floor. Instruction was mostly oral, and the students learned by rote, with the teacher first reciting a phrase and the class then chanting it in unison. Any writing was done Roman style, by scratching a wax tablet with a bone stylus, using the smooth side of the stylus as an eraser, both to correct errors and to clean the surface for further use. Almost certainly, Leonardo would have found computation using Roman numerals tedious. Arithmetic, particularly multiplication by repeated addition, was more speedily done with a counting board (abacus).

After Leonardo had finished his preliminary instruction, his further education in mathematical matters would likely have been in a *fondaco* run by one of his father's friends, where he would have learned the systems of measurement and money and the use of an abacus. The *fondaco*—the name derived from the Arabic *funduq*—was a business establishment in whose front customers and merchants would discuss merchandise, prices, and politics, while in the rear the bookkeeper kept the accounts. Some of the larger ones provided traveling merchants with accommodation and a place to store goods. They were also where government taxes were levied.

The Italian monetary system at the time was like the system used in England until 1970. The lowest-denomination coins were *denari* (pennies), twelve *denari* made a *solidus* (shilling), and twenty *solidi* made a *libra* (pound). The bookkeeper's abacus was divided into seven horizontal rows. The bottom row was used for *denari*, the next one up represented *solidi*, the third was for *librae*, the fourth denoted multiples of twenty *librae*, the fifth hundreds of *librae*, the sixth thousands, and a counter in the seventh and final row represented ten thousand *librae*. Some bookkeepers used different colored counters to indicate in-between numbers; for example, a red counter in the *libra* column might have represented five *librae*, with a black counter being used for a single *libra*.

Young people became bookkeepers by serving an apprenticeship. At first they would simply stand and watch the master at work, then they would graduate to standing alongside the expert, handing him the counters from bowls placed next to the counting board, and finally they would reach the stage when, under the watchful eye of the master, they could perform some of the computations themselves. Yet even in the hands of an expert, the counting board, like any form of abacus, was cumbersome and provided no permanent record of the calculation.

When he was about fourteen years of age, Leonardo would have left the *fondaco* and most likely traveled with an older merchant, a form of apprenticeship system common in those days. Around that time his father summoned him to Bugia. No one knows exactly when he made this voyage. In the introduction to *Liber abbaci*, he later wrote: "When my father, who had been appointed by his country as public notary in the customs at Bugia acting for the Pisan merchants going there, was in charge, he summoned me to him while I was still a child, and having an eye to usefulness and future convenience, desired me to stay there and receive instruction in the school of accounting."

Bugia had originally been a minor Roman colony called Saldae. In the eleventh century, the Berbers revived it, and it rapidly grew to be one of the most important Islamic ports on the Barbary Coast. In the middle of the twelfth century, the Pisans ousted their Genoese rivals and established their own trading port there. Through Bugia, the Pisans exported European goods to North Africa and brought to Europe various Eastern luxury items, including silks, spices, and two commodities the city was particularly famous for: a fine grade of beeswax and a high-quality leather.

Leonardo's father left Pisa to take up his diplomatic post in Bugia sometime between 1180 and 1185, and most likely sent for his son a year or so after his arrival. Leonardo would have begun his journey from Porto Pisano, in all probability setting sail in the spring or early summer. Few vessels put to sea during the fall or winter, when a severe Mediterranean storm could make any voyage hazardous. Pisan ships that made port in Spain or Africa in the fall would have to remain there until the next spring. Ships generally departed on a Monday evening. On the Sunday before leaving, Leonardo probably went to church to pray for a safe journey, followed by a farewell feast with his family. Then, early on the Monday morning, he and the other travelers would have assembled by the Church of Saint Paul on the Arno and mounted horses for the ride to Porto Pisano. Many of his fellow passengers would have been pilgrims heading for the Holy Land.

Sea journeys were always risky. In addition to the possibility of running into a storm—even in the summer—there was always a chance of being attacked by pirates. The Mediterranean swarmed with privateers, originating both from Muslim North Africa—the infamous Barbary Coast—as well as from the Italian ports of Venice, Genoa, and Pisa itself. Italian privateers were financed by groups of shareholders, just as the peaceful merchant vessels were. Although shareholders stipulated that privateers should attack only vessels of enemy countries, once they were on the high seas many captains succumbed to the temptation of a healthy profit and overlooked such restrictions.

Italian merchants had recently started to take advantage of a novel scheme to protect themselves against possible losses of a vessel and its cargo. For a price, a group of wealthy investors would promise to cover any financial loss. The origin of insurance protection in twelfth-century Italy is reflected in our use of the word "policy", which comes directly from the Italian word *polizza*, meaning "promise".

Tunis was the ultimate destination of most voyages. A few ships made the journey by sailing down the Italian coast and then directly across the Mediterranean, but most took a circuitous route that offered the greatest opportunities for trading: first, west to Spain, then south across the Straits of Gibraltar to Morocco, then eastward toward Tunis along the North Africa coast. Each ship's itinerary was fixed in advance by agreement among the merchants whose goods were being carried. Most likely Leonardo's ship took the circuitous route. The two-thousand-mile (ca. three-thousand-kilometer) journey to Bugia would have taken approximately two months. For much of the journey the ship would have stayed close to land. Not only did this make navigation simpler and more reliable; there was a greater margin of safety. If a storm struck, the captain could take his vessel close in to shore for shelter.

The city Leonardo arrived in was one of the most important ports in North Africa, and its Arab traders ventured even farther afield than the Italians, journeying not only around the Mediterranean but to Russia, India, and China, and deep into the interior of Africa. It was part of the Maghreb, a region in North Africa that today comprises Morocco, Algeria, Tunisia, Libya, and Mauritania, but in Leonardo's times the name referred to the much smaller part of that region lying between the high Atlas Mountains and the Mediterranean Sea. The Maghreb was united as a single political entity during the first years of Arab rule, in the early eighth century, and again for several decades under the Berber Almohads from 1159 to 1229. At other times the ties had been primarily through trade and cultural exchange.

Leonardo's father would most likely have lived in the sizable Italian community near the harbor. Most of the business activity was centered on the *fondaco*. The Pisans had signed treaties with the various cities they traded with, governing issues of legal jurisdiction, safe conduct, and access to and from the *fondaco*, the first of which had been signed on July 2, 1133, with Alibibn Yusof, the king of Morocco and Tlemcen. Since Guilielmo brought his son over to prepare him for his future, we can be sure that he encouraged Leonardo to spend a lot of his time in the *fondaco*. The Arabs viewed mathematics in a very practical manner, as something to be used by traders, land surveyors, and engineers, and wrote texts for those professional people, so Guilielmo could well have seen the Hindu-Arabic system as a powerful new tool that would benefit his son.

Much of what we know about Leonardo's time in Bugia comes from the brief prologue with which be began *Liber abbaci*. The first part describes the approach his book takes.*

> You, my Master Michael Scott,† most great philosopher, wrote to my Lord‡ about the book on numbers which some time ago I composed and transcribed to you;§ whence complying with your criticism, your more subtle examining circumspection, to the honor of you and many others I with advantage corrected this work. In this rectification I added certain necessities, and I deleted certain superfluities. In it I presented a full instruction on numbers close to the method of the Indians, whose outstanding method I chose for this science. And because arithmetic science and geometric science are connected, and support one another, the full knowledge of numbers cannot be presented without encountering some geometry, or without seeing that operating in this way on numbers is close to geometry; the method is full of many proofs and demonstrations which are made with geometric figures.[2] And truly in another book that I composed on the

* The very literal translation is by Laurence M. Sigler, 2001. It is based on the second edition of *Liber abbaci*, published in 1228. Hence Leonardo's reference to having "corrected" his work.

† Michael Scott was the philosopher in Frederick II's court.

‡ Leonardo was referring to Frederick.

§ Leonardo was referring to the first edition of *Liber abbaci*.

> practice of geometry* I explained this and many other things pertinent to geometry, each subject to appropriate proof. To be sure, this book looks more to theory than to practice.† Hence, whoever would wish to know well the practice of this science ought eagerly to busy himself with continuous use and enduring exercise in practice, for science by practice turns into habit; memory and even perception correlate with the hands and figures, which as an impulse and breath in one and the same instant, almost the same, go naturally together for all; and thus will be made a student of habit; following by degrees he will be able easily to attain this to perfection. And to reveal more easily the theory I separated this book into xv chapters,‡ as whoever will wish to read this book can easily discover. Further, if in this work is found insufficiency or defect, I submit it to your correction.

At this point, the prologue changes direction, as Leonardo recounted how he came to learn this remarkable new calculating method, thereby providing the only autobiographical information we have about its author. Why he included this is unknown. Like mathematicians before and after him, Leonardo cared little for the history of the discipline. Mathematics is

* Leonardo was referring to his book *De practica geometrie*, which he published in 1220, long after the first edition of *Liber abbaci* had been completed.

† While no doubt true compared to earlier Arabic texts, it does not look at all like this to a modern reader.

‡ Leonardo used Roman numerals here, perhaps because he knew his readers would not understand Hindu-Arabic numerals until they were well past the introduction.

eternal, and exactly when something new is discovered and by whom is of secondary importance. Mathematicians admire those who make great discoveries, but their interest is generally in what is discovered, not in who got there first. Nevertheless, Leonardo presumably realized that the invention his book described was a monumental one, and at the back of his mind may have lurked the notion that one day people would wonder how this great Hindu invention found its way from the Muslim scholars and merchants who had held it for many centuries to the practical trading men of northern Europe. In any event, he broke with tradition and inserted an all-too-brief summary of the part he played in the story.

> As my father was a public official away from our homeland in the Bugia customshouse established for the Pisan merchants who frequently gathered there, he had me in my youth brought to him, looking to find for me a useful and comfortable future; there he wanted me in the study of mathematics and to be taught for some days. There from a marvelous instruction in the art of the nine Indian figures, the introduction and knowledge of the art pleased me so much above all else, and I learned from them, whoever was learned in it, from nearby Egypt, Syria, Greece, Sicily, and Provence, and their various methods, to which locations of business I traveled considerably afterwards for much study, and I learned from the assembled disputations. But this, on the whole, the algorithm and even the Pythagorean arcs, I still reckoned almost an error compared to the Indian method.* Therefore strictly embracing the

* Recall that Gerbert (ca. 980) used the Hindu-Arabic numerals on counters as part of a primitive form of abacus, on which triples of columns were

> Indian method, and attentive to the study of it, from mine own sense adding some, and some more still from the subtle geometric art, applying the sum that I was able to perceive to this book, I worked to put it together in xv distinct chapters, showing certain proof for almost everything that I put in, so that further, this method perfected above the rest, this science is instructed to the eager, and to the Italian people above all others, who up to now are found without a minimum.* If, by chance, something less or more proper or necessary I omitted, your indulgence for me is entreated, as there is no one who is without fault, and in all things is altogether circumspect.

AS PISA'S TRADING representative in Bugia, Leonardo's father would have had the task of maintaining relations with the Muslim authorities, safeguarding the rights of the *fondaco*, keeping records of the goods passing through, and overseeing the proper levying of taxes—activities that would surely have required that Guilielmo was fluent in Arabic. This supposition is borne out by a surviving account from the *funduq* in Damascus in 1183 that refers to the "Christian clerks of the customs" who "write in Arabic, which they also speak."[3] It is reasonable to assume the same was true elsewhere in the Arabic-speaking regions where the Italians did business. In bringing his son to

---

marked with an arc. These were called Pythagorean arcs. When he wrote numbers, Leonardo followed the system of triples, just as we do today when we write numbers like 1,395,281. Leonardo told his readers that even with various enhancements, abacus methods were no match for Hindu-Arabic arithmetic.

* A more colloquial translation of this last clause would be: "Who up to now have lacked this knowledge."

join him to complete his professional education, Guilielmo presumably intended that Leonardo not only learn the Arabs' marvelous new ways of doing arithmetic but also master their language.

There is no way to know for certain whether Leonardo actually did learn to read Arabic, but the evidence suggests so, and this is the accepted view of historians today.[4] With a mastery of Arabic, Leonardo would have been able to broaden his mathematical knowledge well beyond what he could observe in the Bugian marketplace. Among the Arabic scholars, teachers, and students who were known to have moved between the cities of the Maghreb in the late eleventh and early twelfth centuries were several mathematicians: al-Hassār moved from Ceuta to Marrakech and then to Spain, ibn 'Aqnū n moved from Marrakech to Seville and then back to Marrakech, ibn al-Mun'im was born in al-Andalus but worked in Marrakech, and the Andalusian al-Qurashi worked first in Seville and then in Bugia. As a result their mathematical works, and any they had copies of, were presumably circulating between all six cities and would thus have been available in Bugia, where Leonardo could have had access to them.

CHAPTER 4

Sources

AFTER LEONARDO'S IMPORTANT role in the spread of Hindu-Arabic arithmetic became known in the nineteenth century, scholars began to look for the exact written sources he had consulted in writing *Liber abbaci.* Trying to identify source materials written more than eight hundred years ago is inevitably problematic, since many of them may have been lost. To be sure, more authorative sources were more likely to be copied, which increased the chance of the work's survival, but the fact remains that the most historians can do is identify and study sources, or likely sources, among those works that did survive in one form or another.[1]

What seems certain is that Leonardo consulted many sources to write *Liber abbaci,* both Latin and Arabic. Occasionally a particular source can be identified with some confidence; for example, his notation for ascending continued fractions came from the Maghreb mathematical school. But for the most part, historians can only speculate on what manuscripts he read.

The earliest extant Arabic work on Hindu arithmetic is the *Kitab al-fusul fi'l-hisab al-hindi* (Book of chapters on Hindu arithmetic) of Abu'l-Hasan Ahmad ibn Ibrahim al-Uqlidisi, composed in Damascus in 952–53 CE, but that survives only as a manuscript copy written more than two centuries later, in 1186. In fact, Latin manuscripts provide most of the early examples of place-value numerals usage outside India. Thus some of Leonardo's written sources for Hindu-Arabic arithmetic may have been in Latin, the oldest surviving being a copy of al-Khwārizmī's *Arithmetic*. Another available Latin treatise on the system was *Liber ysagogorum alchorismi*, which may have been written by Adelard of Bath. Unlike the majority of the Italian abbacus books, these earlier works were written by, and to a large extent for, scholars. It is possible that Leonardo read or consulted one, and perhaps several such treatises in preparing his description of Hindu-Arabic arithmetic.

Leonardo most definitely based his treatment of algebra in *Liber abbaci* on al-Khwārizmī's *Algebra*. It may not have been the book from which the young Pisan first learned algebra while he was in North Africa, however, since that work was not available in the Maghreb, despite its wide circulation in al-Andalus. Instead, his first source may have been Abū Kāmil's *Kitāb fīl-jabr wa'l muqābala* (Book on algebra). Nevertheless, it is clear that when Leonardo subsequently wrote the more advanced, algebra sections in *Liber abbaci*, he relied heavily on al-Khwārizmī's masterpiece, almost certainly a Latin translation to which he had access in Italy.

The *Algebra* was translated into Latin by Robert of Chester in 1145, by Gherado of Cremona (arguably the greatest translator of the twelfth century, who lived from 1114 to 1187) around 1150, and by Guglielmo de Lunis around 1250. Gherardo's translation is generally regarded as the best and was the most widely used. He titled it *Liber maumeti filii moysi alchoarismi de algebra et almuchabala*.[2] When the present-day scholar Nobuo Miura compared passages in both *Liber abbaci* and Gherardo of Cremona's Latin translation of *Algebra*, she found that many of the ninety problems in Leonardo's chapter on algebra came directly from al-Khwārizmī's text, demonstrating that Leonardo made use of that particular translation.[3]

One of the difficulties facing the medieval historian is illustrated by the confusion in the literature about al-Khwārizmī's full name. Most present-day sources give it as Abū ʿAbdallāh Muḥammad ibn Mūsā al-Khwārizmī, which can be translated as "Father of 'Abdallāh, Mohammed, son of Moses, native of the town of al-Khwārizmī".* The form parallel to Leonardo Pisano (Leonardo of Pisa) would therefore be Muḥammad al-Khwārizmī (Muhammad of Khwārizmī), and the one parallel to Leonardo filius Bonacci Pisano (Leonardo, son of Bonacci, of Pisa) would be Muḥammad ibn Mūsā al-Khwārizmī (Muhammad, son of Moses, of Khwārizmī). This last is the form most present-day scholars use.

Naming conventions are not the only challenge facing the archivist. There are also references in the literature, both ancient and modern, to Abū Ja'far Muḥammad ibn Mūsā al-Khwārizmī. This could have resulted from an erroneous transcription by a careless inattentive scribe, or perhaps Muḥammad al-Khwārizmī had two children, one called Abdallāh, the other Ja'far. Among the sources who cite Abdallāh rather than Ja'far as the mentioned son is Frederic Rosen, who in 1831 published an English-language translation of al-Khwārizmī's *Algebra*.[4] In his preface, Rosen wrote: "ABU ABDALLAH MOHAMMED BEN MUSA, of Khowarezm, who it appears, from his preface, wrote this Treatise at the command of the Caliph AL MAMUN, was for a long time considered as the original inventor of Algebra." That

* A few sources give a longer version: Abū ʿAbdallāh Muḥammad ibn Mūsā al-Khwārizmī al-Majūsī. The last part, which translates as "the Mazdean", indicates that he was a Zoroastrian, but scholars have doubts about the accuracy of that epithet.

would seem to settle the matter. Rosen explained how the confusion arose. On page xi of his preface, he wrote, of the author of the famous algebra text: "He lived and wrote under the caliphat of AL MAMUN, and must therefore be distinguished from ABU JAFAR MOHAMMED BEN MUSA [whose father, Rosen colorfully tells us, was a bandit], likewise a mathematician and astronomer, who flourished under the Caliph AL MOTADED (who reigned A.H. 279–289, A.D. 892–902)."

Clearly, then, the two names referred to two different people. The other mathematician, Abu Ja'far Muhammad ibn Musa al-Khwārizmī, was one of three brothers, the "Sons of Musa" (Banū Mūsā), the others being named Ahmad and al-Hasan. But with both "Muḥammad al-Khwārizmī"s being mathematicians and astronomers, historians have had to exercise caution when citing the literature—particularly since the "father of" part (Abu 'Abdallah or Abu Ja'far) is not found in most manuscripts.

Another tantalizing puzzle arises from Rosen's remark that al-Khwārizmī "was for a long time considered as the original inventor of Algebra." Rosen's words seem to imply definitive knowledge that the famous Arab author was not the inventor of algebra, and that is indeed the case. On page vii of the preface, Rosen wrote: "From the manner in which our author [al-Khwārizmī], in his preface, speaks of the task he had undertaken, we cannot infer that he claimed to be the inventor. He says that the Caliph AL MAMUN encouraged him to write a popular work on Algebra: an expression which would seem to imply that other treatises were then already extant."

In fact, algebra (as al-Khwārizmī described it in his book) was being transmitted orally and being used by people in their jobs before he or anyone else started to write it down. Several authors wrote books on algebra during the ninth century besides al-Khwārizmī, all having the virtually identical title *Kitāb al-ğabr wa-l-muqābala*. Among them were Abū Hanīfa al-Dīnawarī, Abū Kāmil Shujā ibn Aslam, Abū Muḥammad al-'Adlī, Abū Yūsuf al-Miṣṣīṣī, 'Abd al-Hamīd ibn Turk, Sind ibn 'Alī, Sahl ibn Bišr, and Šarafaddīn al-Tūsī.

Al-Khwārizmī's remark, as reported by Rosen in his preface, also states that al-Khwārizmī wrote his algebra book as "a popular work", aimed at a much wider audience than just his fellow scholars. It is full of examples and applications to a wide range of numerical problems dealing with trade, surveying, and the highly complex issues of Islamic legal inheritance. Such a strong focus on applications was typical of Arabic algebra at the time.

The extent to which al-Khwārizmī and Leonardo filled their books with practical examples is not the only similarity between the two authors; another is the frustrating paucity of information about each. As Rosen wrote of al-Khwārizmī, "Besides the few facts which have already been mentioned in the course of this preface, little or nothing is known of our Author's life."[5]

Still another similarity between them is the uncertainty that has surrounded both their names. Although the confusion about al-Khwārizmī's full name has finally been resolved—though the incorrect version continues to appear—for Leonardo, the intended meaning of that appended name "Bigollo" (see page 13) remains something of a mystery.

There is some disagreement as to al-Khwārizmī's mathematical abilities. Did he have creative mathematical talent, or did he merely assemble and transcribe the works of others? Contemporary authorities disagree, saying variously:

> [He was] the greatest mathematician of the time, and if one takes all the circumstances into account, one of the greatest of all time.

> [Al-Khwārizmī] may not have been very original.
>
> It is impossible to overstress the originality of the conception and style of al-Khwarizmī's algebra.
>
> Al-Khwarizmī's scientific achievements were at best mediocre.[6]

None of these commentators argue that al-Khwārizmī's two mathematics books were not hugely important. The disagreement is over his abilities as an original mathematician.[7] In any event, regardless of how good al-Khwārizmī was at producing original mathematics, regardless of which of al-Khwārizmī's books Leonardo consulted and to what extent, regardless of which works by others Leonardo consulted, and regardless of which other scholars Leonardo talked to—all factors of which we have little or no knowledge—what is beyond doubt is that the famous Pisan was a clear beneficiary of the work of al-Khwārizmī.

In his introduction to *Algebra*, al-Khwārizmī stated that the purpose of the book was to explain "what is easiest and most useful in arithmetic, such as men constantly require in cases of inheritance, legacies, partition, lawsuits, and trade, and in all their dealings with one another, or where the measuring of lands, the digging of canals, geometrical computations, and other objects of various sorts and kinds are concerned." He divided the text into three sections: the first part devoted to algebra, giving various rules together with thirty-nine worked problems, all abstract;[8] then a short section on the Rule of Three (see page 74) and mensuration, in which two mensuration problems are solved with algebra; finally, a long section on inheritance problems solved by algebra.

The book begins with an observation about numbers that

seems trivial to modern readers but was profound in al-Khwārizmī's time:

> When I consider what people generally want in calculating, I found that it always is a number. I also observed that every number is composed of units, and that any number may be divided into units. Moreover, I found that every number which may be expressed from one to ten, surpasses the preceding by one unit: afterwards the ten is doubled or tripled just as before the units were: thus arise twenty, thirty, etc. until a hundred: then the hundred is doubled and tripled in the same manner as the units and the tens, up to a thousand; . . . so forth to the utmost limit of numeration.

Understanding what al-Khwārizmī meant requires an appreciation that in his day numbers were regarded as different from quantities of length, a distinction still made in the seventeenth century when Newton invented calculus. The great Arabic mathematician was actually making an uncannily accurate prediction about the degree to which numbers would come to dominate mathematics.

The two words *al-jabr* and *al-muqabala* in al-Khwārizmī's title refer to two steps in the simplification of equations. *Al-jabr* means "restoration" or "completion", that is, removing negative terms, by transposing them to the other side of the equation to make them positive. For example, using one of al-Khwārizmī's own examples, but expressing it with modern symbolic notation), *al-jabr* transforms

$$x^2 = 40x - 4x^2$$

into

$$5x^2 = 40x.$$

Al-muqabala means "balancing" and is the process of eliminating identical quantities from the two sides of the equation. For example (again in modern notation), one application of *al-muqabala* reduces

$$50 + 3x + x^2 = 29 + 10x$$

to

$$21 + 3x + x^2 = 10x$$

and a second application reduces that to

$$21 + x^2 = 7x.$$

These are the methods we use today to simplify and hence solve equations, which explains why a meaningful, modern English translation for al-Khwārizmī's Arabic book title *Hisâb al-Jabr wa'l-Muqâbala* would be, simply, "Calculation with algebra".

Today we interpret completion and restoration very differently from medieval mathematicians. The Arabs did not acknowledge negative numbers. For instance, they viewed "ten and a thing" $(10 + x)$ as a composite expression that entailed two types of number ("simple numbers" and "roots"), but they did not see "ten less a thing" $(10 - x)$ as composite. Rather, they thought of it as a single quantity, a "diminished" 10, or a 10 with a "defect" of x. The 10 retained its identity, even though x had been taken away from it. Thus, in a rhetorical equation like "ten less a thing equals five things" the "ten less a thing" was viewed as a deficient "ten" which needed to be restored, and the Arabic mathematicians would write "So restore the ten by the thing and add it to the five things" to get the equation "ten equals six things." For confrontation, in an equation like "ten and two things equals six things", they would "confront" the two things with the six things, which entailed taking their difference, to get the equation "ten equals four things".[9]

Al-Khwārizmī, like Leonardo after him, developed his algebra in rhetorical fashion, using words, and would not have understood the symbolic derivations above. Arab mathematicians called the unknown quantity the "thing" (*shay*) or "root" (*jidhr*). The word *jidhr* means "the origin" or "the base", also "the root of a tree", and that may be the origin of our present-day expression "root of an equation". (Our word "root" is a translation of the Latin word *radix*, but its connection to the Arabic is disputed.)

In addition to his two books on mathematics, al-Khwārizmī wrote a revised and completed version of Ptolemy's *Geography*, consisting of a general introduction followed by a list of 2,402 coordinates of cities and other geographic features. He gave his book the title *Kitāb ṣūrat al-Arḍ* (Book on the appearance of the Earth or The image of the Earth) and finished it in 833.* The complete title of the Latin edition translates as "Book of the appearance of the Earth, with its cities, mountains, seas, all the islands and rivers, written by Abu Ja'far Muhammad ibn Musa al-Khwārizmī, according to the geographical treatise written by Ptolemy the Claudian." Once again, that incorrect name *Abu Ja'far* appears. Perhaps the copyist mistook him for Muhammad, one of the Musa brothers. This may in fact be the source of the present-day confusion about the name. The

* There is only one surviving copy, which is kept at the University of Strasbourg Library. There is a Latin translation at the Biblioteca Nacional de España in Madrid.

biographer G. J. Toomer probably consulted that Latin text to write the description of al-Khwārizmī for the *Dictionary of Scientific Biography* (New York, 1970–90), since the entry lists him as *Abu Ja'far Muhammad ibn Musa al-Khwārizmī*, and presumably it is from there that the error propagated through the literature.

AN ORIGINAL WORK written in Latin, Leonardo's *Liber abbaci* was clearly based in part on the earlier writings of al-Khwārizmī and other Arabic mathematicians. Other than his known use of Gherardo's Latin translation of al-Khwārizmī's *Algebra*, however, it is not clear whether Leonardo used Arabic manuscripts or Latin translations, or whether he read them in Bugia, elsewhere in North Africa, or in Italy after his return to Pisa. At that time, many Arabic texts had found their way to Europe, particularly Spain, where Latin translations were made—not just translations of original works by Arab mathematicians but also Arabic translations from the ancient Greek, including Euclid's *Elements* and Ptolemy's *Almagest*.

Much of the translation work was carried out in the area around the cathedral in the Spanish city of Toledo. Though all European scholars of the time knew Latin, few had mastered Arabic, so the translation was often done in two stages. One scholar—often a Jewish or Muslim scholar living in Spain—would make the translation from the Arabic to some common language, and a second scholar would then translate from that language into Latin. In the same way, many ancient Greek texts, from Aristotle to Euclid, were also translated into Latin.

In addition to Gherado of Cremona, who translated al-Khwārizmī's *Algebra*, a colleague called "magister Iohanne", or "magister Iohannes Hispalensis," translated the *Liber alchoarismi de pratica arismetice*, the most complete exposition of

Arabic arithmetic and algebra of the twelfth century. It is likely, though not certain, that the same Iohannes wrote *Liber mahamalet,* an original book on commercial arithmetic based on Arabic material.

Many of the Latin manuscripts produced in Toledo found their way to Italy. So even if Leonardo had no access to a particular Arabic text while on his travels through North Africa, he could have consulted it closer to home. Scholars today seem generally agreed that in writing *De practica geometrie* he made direct use of both Euclid's *Elements* and Plato of Tivoli's *Liber embadorum* (1145), which is based on the second book of al-Khwārizmī's *Algebra.* (Plato is known only through his writings, at least some of which were produced in Barcelona between 1132 and 1146.)

Aside from al-Khwārizmī's two books, there is less agreement about Leonardo's other sources for *Liber abbaci.* One obvious possibility is that he had access to some of the Arabic texts—or Latin translations thereof—written after al-Khwārizmī. In particular, there are parallels between *Liber abbaci* and works of the Egyptian-born Abū Kāmil Shujā' ibn Aslam ibn Muḥammad ibn Shujā (ca. 850–ca. 930).[10] Abū Kāmil's *Algebra* appears to have been a major source for Leonardo's treatment of algebra, not only in *Liber abbaci* but also in his other books *De practica geometrie* and *Flos,* and may have been the source from which he first learned algebra.[11] Abū Kāmil's book has seventy-four worked-out problems, and many of the more complicated ones, with identical solutions, are found in *Liber abbaci.* What is not clear is whether Leonardo used an Arabic text or a Latin translation.

Abū Kāmil was the first major Arabic algebraist after al-Khwārizmī. By all accounts he was a prolific author. There are references to works with the titles *Book of fortune, Book of the key to fortune, Book of the adequate, Book on omens, Book of the kernel*, *Book of the two errors*, and *Book on augmentation and diminution*. None of these have survived. Works that have include the *Book on algebra*, the *Book of rare things in the art of calculation*, and the *Book on surveying and geometry*.

Although al-Khwārizmī's book was primarily intended for practitioners, he included some proofs for those interested in the reasons for some results. In his books, Abū Kāmil extended the range of geometric proofs. He was also the first to work freely with irrational coefficients.

Further advances in algebra were made in the Maghreb in the twelfth to fifteenth centuries, by a highly organized teacher-student network linked to mosque and madrassa teaching. The Maghrebs used abbreviations for both unknowns and their powers and for operations, an innovation that inspired parallel advances in Italian algebra, leading ultimately to the development of modern symbolic algebra.

Since Leonardo's notation for ascending continued fractions comes from the Maghreb mathematical school, he likely had access to some of their writings, either in Arabic or in a Latin translation. It seems clear that he also consulted the *Book on ratio and proportion* of Ahmad ibn Yusuf ibn ad-Daya (Ametus filius Iosephi) and the *Book on geometry* by the Banū Mūsā. He also used problems from the *Liber mahamalet*.

Leonardo may have used other sources, but recent scholarship has ruled out one obvious candidate: he did not have Omar Khayyám's *Algebra* at his disposal.[12] Omar Khayyám is better known in the West today as a poet than a mathematician, but that reflects more on the values of today's Western society than on the inherent merits of Khayyám's work. While his poetry is competent, and liked by many, few would seriously claim it is on a par with the very best. His mathematical work, on the other hand, is first-rate. Born in 1048 in Nichapur, Persia (now Iran), Khayyám died there in 1131. As a young man he studied philosophy; by the time he was twenty-five, he had written books on arithmetic, algebra, and music. In 1070, he moved to Samarkand in Uzbekistan, and there he wrote his great work in algebra, an analysis of polynomial equations, titled *Algebra wa al-muqabala* (Proofs of algebra problems).

There is, then, some uncertainty regarding Leonardo's sources for *Liber abbaci.* Historians have faced another challenge trying to determine the sequence of events that followed the book's publication. In particular what role did *Liber abbaci* play in the arithmetic revolution that swept through Europe after Leonardo completed it. The one thing we know for sure is what Leonardo wrote in *Liber abbaci* itself—and it was considerable.

CHAPTER 5

Liber abbaci

IN ADDITION TO ITS TREATMENT of Hindu-Arabic arithmetic, *Liber abbaci* covers the beginnings of algebra and some applied mathematics. Some of the methods Leonardo described may have been his own invention, but he obtained much from existing sources, primarily Arabic texts or Latin translations thereof, and from discussions with the Arabic mathematicians he encountered on his travels. In all cases, he provided rigorous proofs to justify the methods, in the fashion of the ancient Greeks, and illustrated everything with copious worked examples designed to provide exercises in using the new methods.

Leonardo divided the book into fifteen chapters, the titles of which vary from manuscript to manuscript, suggesting that the scribes who made copies felt free to make what they felt were clarifying improvements. The titles in Sigler's English translation are:

Dedication and prologue

1. On the recognition of the nine Indian figures and how all numbers are written with them; and how the numbers must be held in the hands, and on the introduction to calculations

2. On the multiplication of whole numbers
3. On the addition of them, one to the other
4. On the subtraction of lesser numbers from greater numbers
5. On the divisions of integral numbers
6. On the multiplication of integral numbers with fractions
7. On the addition and subtraction and division of numbers with fractions and the reduction of several parts to a single part
8. On finding the value of merchandise by the Principal Method
9. On the barter of merchandise and similar things
10. On companies and their members
11. On the alloying of monies
12. On the solutions to many posed problems
13. On the method elchataym and how with it nearly all problems of mathematics are solved
14. On finding square and cubic roots, and on the multiplication, division, and subtraction of them, and on the treatment of binomials and apotomes and their roots
15. On pertinent geometric rules and on problems of algebra and almuchabala.

His opening chapter describes how to write—and read—whole numbers in the Hindus' decimal system. Leonardo began: "These are the nine figures of the Indians: 9 8 7 6 5 4 3 2 1. With these nine figures, and with this sign 0 which in Arabic is called zephirum, any number can be written, as will be demonstrated." He went on to explain the principles of place value, describing the forms of the numerals, and showing how to write large numbers (either by putting a dot—*adcentare*—above each hundred, and below each thousand, or by linking groups of three numerals with a small curved stroke called a *virgula*). When his chapter title promised to explain "how the numbers must be held in the hands," he meant that quite literally. He described a procedure to calculate on the fingers, widely used by medieval scholars and traders, which was regarded as the easiest and quickest way of performing calculations. Manuscript copies of *Liber abbaci* and those of many other arithmetic books that were to follow often included a drawing showing the various finger positions used to represent different whole numbers. Leonardo also provided addition and multiplication tables to be referred to—or memorized—in order to facilitate computations. In all, he devoted several pages to this introductory description of the numerals, which would have been his readers' first encounter with modern numbers.

The approach Leonardo took to multiplication in chapter 2 differs little from the one used today to teach children how to multiply two whole numbers together. He began with the multiplication of pairs of two-digit numbers and of multidigit numbers by a one-place number and then worked up to more complicated examples. He described various methods for checking the answers.* It is interesting to note that Leonardo described multiplication before covering addition in chapter 3 and subtraction in chapter 4.†

Whole number division and simple fractions are described in chapter 5. An "integral number", or "integer", is a technical term for a whole number, positive, negative, or zero.

The topic of chapters 6 and 7 is what are today called mixed

* One of them, "casting out nines," was just going out of use in the United Kingdom when I was taught arithmetic there in the 1950s.

† Teachers today usually introduce addition and subtraction before multiplication.

numbers, numbers that comprise both a whole number and a fractional part. Leonardo explained that you calculate with them by first changing them to fractional form (what we would today call "improper fractions"), computing with them, and then converting the answer back to mixed form.

Having described the basic methods of Hindu-Arabic arithmetic in the first seven chapters, Leonardo devoted most of the remainder of the book to practical problems. Chapters 8 and 9 provide dozens of worked examples on buying, selling, and pricing merchandise, using what we would today call reasoning by proportions—the math we use to check the best deal in the supermarket. For example, Leonardo asked: if two pounds of barley cost five *solidi*, how much do seven pounds cost, showing how to work out the answer. In chapter 10, he explained how to use similar methods to manage investments and profits of companies and their members, and showed how to decide who should be paid what.

Chapter 11, "On the alloying of monies", met an important need at that time. Italy had the highest concentration of different currencies in the world, with twenty-eight different cities issuing coins during the course of the Middle Ages, seven in Tuscany alone. Their relative value and the metallic composition of their coinage varied considerably, from one city to the next and over time. This state of affairs meant good business for money changers, and *Liber abbaci* provided them with plenty of examples on problems of that nature. Also, with governments regularly revaluing their currencies, gold and silver coins provided a more stable base, and since most silver coinage of the time was alloyed with copper, problems about minting and alloying of money were important.

The lengthy chapter 12 presents 259 worked examples, some requiring only a few lines to solve, others spread over several densely packed pages. In modern terminology, the main focus is essentially "algebra"—not the symbolic reasoning we associate with the word today, rather algebraic reasoning expressed in ordinary language, often referred to as "rhetorical algebra". Symbolic algebra did not appear until 1591, when the French amateur mathematician and astronomer François Viète published his book *In artem analyticam isagoge* (Introduction to the analytic art), explaining how to formulate and solve an equation in symbolic form, much as we do today. Mathematicians in Leonardo's time used a variey of methods to solve problems that today would be handled using symbolic algebra.

The term *elchataym* that appears in the title of chapter 13 refers to a rule known also as "Double False Position". The word is Leonardo's Latin transliteration of the Arabic *al-khata'ayn*, which means "the two errors". The name reflects the fact that you start with two approximations to the sought-after answer, ideally one too low, the other too high, and then reason to adjust them to give the correct answer. Leonardo formulated his worked problems in several ingenious ways, in terms of snakes, four-legged animals, eggs, business ventures, ships, vats full of liquid which empty through holes, how a group of men should share out the proceeds when they find a purse or purses, subject to various conditions, how a group of men should each contribute to the cost of buying a horse, again under various conditions, as well as some problems in pure number terms.

Though most of the problems in chapters 12 and 13 are of a practical nature, Leonardo included some that have a whimsical flavor. For example, if a spider climbs a wall of a cistern, advancing a certain number of feet each day, but slips back so many feet each night, how long will it take to climb out? For some of these entertaining problems he presented clever solutions that may have been of his own devising.

In his penultimate chapter, Leonardo's main focus was on methods for handling roots. He used the classifications given by Euclid in book 10 of *Elements* for the sums and differences of unlike roots—known as *binomials* and *apotomes* respectively. The discovery that $\sqrt{2}$ is irrational led the ancient Greeks to a study of what they called "incommensurable magnitudes". Euclid's term for a sum of two incommensurables, such as $\sqrt{2}+1$, was *binomial* (a "two-name" magnitude), and a difference, such as $\sqrt{2}-1$, he called an *apotome*. Handling incommensurables by means of what we would now regard as algebraic expressions was a common feature of Greek and medieval mathematics, and was the way they dealt with algebraic binomials such as "two things less a dirham" ($2x-1$ in modern algebraic notation). This is one reason why many medieval Italian treatments of algebra began with a long chapter on binomials and apotomes.

Chapter 15, the final chapter, deals with algebraic equations—again expressed in rhetorical fashion—using the methods descibed in al-Khwārizmī's *Algebra*. Leonardo also fulfilled a promise he made in the preface, to deal with things "pertaining to geometry." This is not a treatment of geometry as such, but rather an investigation of the connection between geometry and arithmetic, particularly the use of algebraic and geometric ideas to solve problems in arithmetic.

FOR THE TWENTY-FIRST-CENTURY reader who picks up the English translation of *Liber abbaci*[1] and starts to read it, its most striking feature is the enormous number of worked examples that fill the book's six hundred pages. A beginning math student today does not require, and would not tolerate, such a deluge of examples on basic arithmetic and algebra, but Leonardo was writing for a thirteenth-century audience. To most people at the time, symbolic numbers and their arithmetic were alien. Arithmetic was something practiced by the merchants, who used finger reckoning or an abacus.*

Many of the problems Leonardo presented involved weights. This was an important but complicated subject, since units of weight differed from one city to another and therefore frequently had to be converted. For example, one worked problem in chapter 8 is titled: "On finding the worth of Florentine rolls† when the worth of those of Genoa is known".[2] A typical worked problem in this section of the book starts:

> If one hundredweight of linen or some other merchandise is sold near Syria or Alexandria for 4 Saracen bezants, and you will wish to know how much 37 rolls are worth, then . . . [3]

In chapter 10, "On companies and their members", Leonardo demonstrated valuable methods for solving problems such as determining the payouts in the following scenario:

> Three men made a company in which the first man put 17 pounds, the second 29 pounds, the third 42 pounds, and the profit was 100 pounds.[4]

Toward the end of chapter 11, he gave a curious problem that became quite well known to mathematicians. It is called "Fibonacci's Problem of the Birds":

---

* They particularly liked finger reckoning because it allowed them to perform their calculations on the spot, without the need for any apparatus. Given the simplicity of most of their computations, it was an adequate method.

† The "roll" was a unit of weight, equal to twelve ounces.

ON A MAN WHO BUYS THIRTY BIRDS OF THREE KINDS FOR 30 DENARI

> A certain man buys 30 birds which are partridges, pigeons, and sparrows, for 30 denari. A partridge he buys for 3 denari, a pigeon for 2 denari, and 2 sparrows for 1 denaro, namely 1 sparrow for ½ denaro. It is sought how many birds he buys of each kind.[5]

What makes this problem particularly intriguing is that apparently there is not enough information to solve it. If you let x be the number of partridges, y the number of pigeons, and z the number of sparrows, then the information you are given leads to two equations:

$$x + y + z = 30 \quad \text{(the number of birds bought equals 30)}$$
$$3x + 2y + \tfrac{1}{2}z = 30 \quad \text{(the total price paid equals 30)}$$

But in general three equations are needed to find three unknowns. What makes this problem different is the availability of one crucial additional piece of information that enables the solution: the values of the three unknowns must all be positive whole numbers. (The man buys three kinds of birds, so none of the unknowns can be zero, and he surely does not buy fractions of birds.) Start by doubling every term in the second equation to get rid of the fraction:

$$x + y + z = 30$$
$$6x + 4y + z = 60$$

Subtract the first equation from the second to eliminate z:

$$5x + 3y = 30$$

Since 5 divides the first term and the third, it must also divide y. So y is one of 5, 10, 15, et cetera. But y cannot be 10 or anything bigger, since then it could not satisfy that last equation. Thus $y=5$. It follows that $x=3$ and $z=22$. Leonardo, as usual, presented the solution in words, not symbols, but apart from that, this is his solution.

Like mathematics teachers and authors before him and since, Leonardo clearly knew that many of the people who sought to learn from him would have little interest in theoretical, abstract problems. Though mathematicians feel entirely at home in a mental world of abstract ideas, most people prefer the more concrete and familiar. And so, in order to explain how to use the new methods he learned during his visit to North Africa, Leonardo looked for ways to dress up the abstractions in familiar, everyday clothing. The result is a class of problems that today go under the name "recreational mathematics".

For instance, Leonardo presented a series of "purse problems" to try to put into everyday terms the mathematical challenge associated with dividing up an amount of money—or anything else that people may have wanted to share—according to certain rules. The first was:

> Two men who had denari found a purse with denari in it; thus found, the first man said to the second, If I take these denari of the purse, then with the denari I have I shall have three times as many as you have. Alternately the other man responded, And if I shall have the denari of the purse with my denari, then I shall have four times as many denari as you have. It is sought how many denari each has, and how many denari they found in the purse.[6]

Students today would be expected to solve this problem using elementary algebra (equations), and this takes at most a few lines. But restricted to the methods available at the time, Leonardo's solution filled almost half a parchment page. (At heart, it is the same as the modern answer by algebra, but without symbolic equations it takes a lot more effort, and a great deal more space on the page, to reach the answer.)

More complicated variations followed, totaling eighteen purse problems in all, including a purse found by three men, a purse found by four men, and finally a purse found by five men. Each problem took a full parchment sheet to solve. Adding still more complexity, Leonardo presented a particularly challenging problem in which four men with *denari* find four purses of *denari,* the solution of which took several pages.

Although many of the variants to the purse problem Leonardo presented seem to be of his own devising, the original problem predated him by at least four hundred years. In his book *Ganita Sara Sangraha*, the ninth-century Jain mathematician Mahavira (ca. 800–870) presented his readers with this problem:

> Three merchants find a purse lying in the road. The first asserts that the discovery would make him twice as wealthy as the other two combined. The second claims his wealth would triple if he kept the purse, and the third claims his wealth would increase five fold.

The reader has to determine how much each merchant has and how much is in the purse. This is precisely Leonardo's first purse problem in chapter 12 of *Liber abbaci.* Presumably Leonardo came across the puzzle by way of an Arab text.

Leonardo's purse problems involved divisions that required only whole numbers. To explain how to proceed when fractions are involved, he used a different scenario his readers could relate to: buying horses. The first horse problem reads:

> Two men having bezants found a horse for sale; as they wished to buy him, the first said to the second, If you will give me ⅓ of your bezants, then I shall have the price of the horse. And the other man proposed to have similarly the price of the horse if he takes ¼ of the first's bezants. The price of the horse and the bezants of each man are sought.[7]

Again, today we would solve this problem using symbolic algebraic equations. Leonardo solved it thus:

> You put ¼ ⅓ in order, and you subtract the 1 which is over the 3 from the 3 itself; there remains 2 that you multiply by the 4; there will be 8 bezants, and the first has this many. Also the 1 which is over the 4 is subtracted from the 4; there remains 3 that you multiply by the 3; there remains 9 bezants, and the other man has this many. Again you multiply the 3 by the 4; there will be 12 from which you take the 1 that comes out of the multiplication of the 1 which is over the 3 by the 1 which is over the 4; there remain 11 bezants for the price of the horse; this method proceeds from the rule of proportion, namely from the finding of the proportion of the bezants of one man to the bezants of the other; the proportion is found thus.[8]

This definitely does not read like a modern-day math textbook. It is far more reminiscent of a cooking recipe written for beginners and therefore leaving nothing to chance. Leonardo explained step-by-step what digits had to be written where, and what to do to them.

Among a total of twenty-nine horse-type problems Leonardo

presented is one in which five men buy five horses[9] and another, which he titled "A Problem Proposed to Us by a Most Learned Master of a Constantinople Mosque",[10] in which five men buy not a horse but a ship, and still another where seven men buy a horse.[11]

One of Leonardo's problems leads to the particularly nasty answer that a certain businessman walks away from a partnership in Constantinople with a profit of

$$\frac{1}{2}\,\frac{7}{8}\,\frac{1}{8}\,\frac{4}{8}\,\frac{0}{8}\,\frac{21169}{24767}\ 206 \text{ bezants}$$

When Europeans in Leonardo's time learned the Hindu-Arabic number system, they wrote fractions before whole numbers, and built those fractions up from right to left, with each new fraction representing that part of what is to the right. For example,

$$\frac{1}{2}\,\frac{2}{3}\,\frac{4}{5} \text{ means } \frac{1}{2\times3\times5}+\frac{2}{3\times5}+\frac{4}{5}, \text{ i.e., } \frac{29}{30}$$

The right-to-left ordering may simply be a carryover from the writing of Arabic, although for the most part Arabic texts expressed Hindu-Arabic numbers rhetorically, using words instead of symbols. Leonardo would have articulated the above fraction as the Arabic mathematicians would both write and speak it: "Four fifths, and two thirds of a fifth, and one half of a third of a fifth."

Decimal expansions are a special case of this notation when the denominators are all 10. For example, Leonardo would have written today's decimal number 3.14159 as

$$\frac{9}{10}\,\frac{5}{10}\,\frac{1}{10}\,\frac{4}{10}\,\frac{1}{10}\,3$$

Though decimal representation seems far simpler to us today, there was little need for it in Leonardo's time, as no one counted anything in tenths. In fact, the method they used to represent fractions was particularly well suited for calculations involving money. In the monetary system used in medieval Pisa, 12 *denari* equaled 1 *solidus* and 20 *solidi* equaled 1 *libra*, so 2 *librae*, 7 *solidi*, and 3 *denari* would be written

$$\frac{3}{12}\frac{7}{20}2$$

Units of weight and measure could be even more complex. According to Leonardo, Pisan hundredweights "have in themselves one hundred parts each of which is called a roll, and each roll contains 12 ounces, and each of which weighs ½ 39 pennyweights; and each pennyweight contains 6 carobs and a carob is 4 grains of corn."[12]

Interestingly, an Arabic arithmetic text written by al-Uqlidisi in Damascus in 952 did in fact use place-value decimals to the right of a decimal point, but no one saw any particular reason to adopt it, and so the idea died, not to reappear again for five hundred years, when Arab scholars picked up the idea once more. Decimal fractions were not used in Europe until the sixteenth century.

Fractions written after the whole number part in Leonardo's time denoted multiplication. For example, ½ of 3.14159 would have been written

$$\frac{9}{10}\frac{5}{10}\frac{1}{10}\frac{4}{10}\frac{1}{10}3\frac{1}{2}$$

THOUGH LEONARDO DIVIDED up his book in terms of different kinds of applications, underlying his solutions were a surprisingly small number of different techniques. The most notable were "the Rule of Three", "the Rule of False Position", "the Rule of Double False Position", and a nonsymbolic form of "Algebra", in particular a method he called *regula recta* (Direct Method), which we would describe today as rhetorical algebra restricted to linear equations.

The Rule of Three was an ancient method for solving proportions problems, such as:

> If 10 men can dig a trench in 4 days, how long will 7 men take to dig a similar trench?

Today we solve such problems fairly easily. One approach is to calculate the amount of time it would take one man to dig a trench, namely $4 \times 10 = 40$ days (since 10 men would dig ten times as fast), and then divide by 7 to get the time it would take 7 men (who would progress seven times as fast), namely $40/7 = 5\frac{5}{7}$ days.

In general, the Rule of Three can be applied to problems that can be described in terms of proportions, like this:

> Given three numbers, find a fourth in such proportion to the third as the second is to the first.

And the rule then says:

> First place the numbers in such order that the first and third be of one kind, and the second the same as the number required. Multiply the second and third numbers together, and divide the product by the first, the quotient will be the answer to the question.

In modern algebraic terms, given numbers a, b, c, we have to find a number x such that x is to c as b is to a, i.e., so that

$$x : c = b : a$$

To solve this problem, you multiply both sides by c to get

$$x = b \cdot c / a$$

The equation $a \cdot x = b \cdot c$ that you get by multiplying the original proportion by $a \cdot c$ was sometimes referred to as "the product of the means equals the product of the extremes".

In the case of the trench problem, $a=7$, $b=4$, $c=10$ (so the first and the third are of one kind, namely number of men, and the second is of the same kind as the number required, namely number of days), and the rule says the answer is given by multiplying 4 by 10, to give 40, and then dividing by 7, to give 5⁵⁄₇ days.

The Rule of Three was known in China as early as the first century. Indian texts discussed it from the fifth century onward—it appeared in the *Āryabhatīya* and was extended and elaborated upon in Bhāskara's commentaries, in which he applied it to problems quite similar to those analyzed by Leonardo. It was introduced into the Islamic world in about the eighth century. Renaissance Europe called it the Golden Rule, presumably because of the importance in commerce of solving relative proportions problems (which arise all the time in pricing and exchange) and the consequent utility of having a simple mechanical procedure that anyone could use. It was the basis for two of the other rules that were in common use at the time: the Rule of False Position and the Rule of Double False Position.

In modern terms, the Rule of False Position solves simple linear equations of the type $Ax = B$. Used by the ancient Egyptians, it consists of making a guess at the solution and then finding the correct answer by forming a proportion with the desired result:

result from guess : target value = guess : solution

For example, we can use the method of False Position to solve the following problem from the famous Rhind Papyrus, currently housed in the British Museum:

> Find a quantity such that when its seventh part is added to it gives 19.

In this problem, the target value is 19. We must start with a guess for the answer. It doesn't really matter what we choose, but it is best to select something that makes the arithmetic easier. Since the problem involves the seventh part of the answer, a good guess is 7. If we make that guess, then adding a seventh to it gives $7+1=8$. That is the result from the guess. So, according to the rule:

$$8 : 19 = 7 : solution$$

By the Rule of Three, you get *solution* by multiplying 19 by 7 and dividing by 8, namely $^{133}/_{8}$, or $16\frac{5}{8}$.

Leonardo devoted most of his lengthy chapter 12 to solving problems using the Rule of False Position. He divided it into nine parts, each devoted to problems of a particular kind. The third part was titled "Problems of trees, and other similar problems for which solutions are found", and contains almost a hundred solved examples. The first five problems established what Leonardo meant by a tree problem: One wants to know the total length of a tree when given the proportion that lies beneath the ground. Presenting such problems in terms of trees was simply a convenient artifact; many practical problems that arise in commerce require finding the total amount based on a known proportion. It is because Leonardo solved such problems using the Rule of False Position that he called the method the "Tree Rule". The first problem says:

> There is a tree ⅓ and ¼ of which lies under the ground, and it is 21 palms. It is sought what is the length of the tree.[13]

The expression "⅓ and ¼" means "⅓ + ¼"; expressing fractional parts using sums of reciprocals was common practice at the time. Leonardo first solved the problem using the Rule of Three he discussed earlier:

> Because ¼ and ⅓ are found in 12, you understand the tree itself to be divided into 12 equal parts, of which a third and a fourth, that is 7 parts, are 21 palms. Therefore proportionally, as the 7 is to the 21, so the 12 parts to the length of the tree. And because whenever four numbers are proportional, the product of the first by the fourth is equal to the product of the second by the third, therefore if you multiply the second 21 by the known third 12, and you divide by the first number similarly, namely by the 7, there will come up 36 for the fourth unknown number, namely for the length of that tree.

Next he introduced his Tree Rule (i.e., the Method of False Position):

> There is indeed another method which we use, namely that you put for the unknown thing some arbitrary known number which is divided evenly by the fractions which are put in the problem itself. And according to the posing of that problem, with that put number you strive to discover the proportion occurring in the solution of that problem.

Leonardo then used this rule to obtain his alternative solution:

> For example, the sought number of this problem is the length of the tree. Therefore you put that to be 12, since it is divided evenly by the 3 and the 4 which are under the fraction lines. And because it is said ⅓ and ¼ of the tree are 21, you take ⅓ + ¼ of the put 12. There will be 7, which if it were by chance 21 we would certainly have what was proposed, namely that the tree be 12 palms. But because 7 is not 21, it turns out therefore, proportionally as the 7 is to the 21 so the put tree to the sought one, namely the 12 to the 36. Therefore one is in the habit of saying, "For the 12 which I put, 7 results. What should I put so that 21 results?" And if it is said this way, the outermost numbers should be multiplied together, namely the 12 by the 21, and the result should be divided by the remaining number.

In terms of the arithmetic needed, Leonardo's two methods are the same. But they are conceptually different, and Leonardo drew a clear distinction between them. The first method, the Rule of Three, uses a convenient subdivision of the unknown; the second method, the Tree Rule, involves guessing a value for the unknown and then correcting it by a proportion calculation.

The Rule of Double False Position, the focus of Leonardo's chapter 13, seems to have its origins in China sometime before 100 C.E., where it was called *Ying pu tsu* (too much and not enough) or sometimes the *thia nu* rule, *thia* meaning "the latest appearance of the waning moon", *nu* "the earliest appearance of the waxing moon". It was known to the Arabic mathematicians, who called it *hisab al-khata'ayn*, and was described in the works of al-Khwārizmī in the ninth century. Medieval European scholars called it *elchataym* or *el cataym*. It can be used to solve linear equations not only of the form $Ax = B$, for which the Method of False Position can be used, but also the more general form $Ax + B = C$.

The rule begins with two guesses to the answer and then uses a systematic method (or perhaps inspired guesswork) to refine them. A simple application would be to solve a problem like this:

> A, B, and C built a house which cost $500, of which A paid a certain sum, B paid $10 more than A, and C paid as much as A and B both; how much did each man pay?

With such simple numbers, we could solve this problem by guesswork. A solution by the Method of Double False Position would go like this: We begin with a low guess, say A pays $80. Then B pays $90 and C pays $170. This means that altogether the three pay $340. This answer has an error of $500 - 340 = 160$ dollars. Now we start with a high guess, say A pays $150. Then B pays $160 and C pays $310, giving a total of $620 for all three. This time the error is $500 - 650 = -120$ dollars. Expressed in modern terms as a formula, the Rule of Double False Position says that the correct answer for A is given by

$$\frac{\text{(first guess)(second error)} - \text{(second guess)(first error)}}{\text{(second error)} - \text{(first error)}}$$

This gives:

$$\frac{(80)(-120) - (150)(160)}{(-120) - (160)} = \frac{-9600 - 24000}{-280} = 120$$

Thus A pays $120, B pays $130, and C pays $250.

A modern solution would use algebra, formulating three equations that represent the information given:

$$A + B + C = 500, \quad B = A + 10, \quad C = A + B$$

Three equations, three unknowns, which you solve by substitution to give A = $120, B = $130, C = $250.

Leonardo used the rule to solve the following problem in chapter 13 of *Liber abbaci*:

> A certain man went on business to Lucca, next to Florence, and then back to Pisa, and he made double in each city, and in each city he spent 12 denari, and in the end nothing was left for him. It is sought how much he had at the beginning.[14]

He calculated as follows:

> You indeed put it that he has 12 denari of which he made double in the first trip, and thus he had 24 denari from which he spent 12 denari and there remained for him another 12 denari, of this he made double in the remaining two trips and he spent in each 12 denari; there remained for him at the end 12 denari. Therefore in the position I erred in value by plus 12; therefore you put it that he had 11 denari from which, as he made double in the three trips and spent in each 12 denari, there remained for him at the end 4 denari, namely 8 fewer than in the first position. And therefore this position is too large. Whence you say: for the 1 which I decreased in the capital I approximated more closely by 8; how many shall I decrease again so that the approximation is decreased by 4 further? You therefore multiply the 4 by the 1, and you divide by the 8, the quotient will be ½ of one denaro; this subtracted from the 11 denari leaves 10½ denari for the capital.

In modern terminology, Leonardo's first guess was 12 with the resulting first error 12, and his second guess was 11 with the second error being 4, so the formula says the correct answer is

$$\frac{12 \times 4 - 11 \times 12}{4 - 12} = \frac{-84}{-8} = 10\frac{1}{2}$$

The formula—that mysterious-looking rule—is simply a modern way to describe the method. Medieval mathematicians would set up a proportion between the two incorrect answers and use the Rule of Three, which is what Leonardo was doing. The simplest solution is the modern one using symbolic algebra. You let A be the amount the man starts out with, then calculate that after visiting Lucca he has $2A - 12$, after Florence $2(2A - 12) - 12 = 4A - 36$, and after returning to Pisa $2(4A - 36) - 12 = 8A - 84$. This final amount must be 0, so $8A = 84$, giving $A = 10\frac{1}{2}$.

The algebraic method was known to Leonardo, indeed to the Arab scholars centuries before him, and was described by al-Khwārizmī around 830 in his algebra book. Without symbolic notation, algebraic solutions were not significantly shorter than the other methods he described; nevertheless, starting in the third section of chapter 12 of *Liber abbaci*, Leonardo gave many examples of how to solve problems using algebra. He introduced his alternative approach with a puzzle about money:

> [I]t is proposed that one man takes 7 denari from the other, and he will have five times the second man. And the second man takes 5 denari from the first, and he will have seven times the denari of the first.[15]

The problem is to determine how much each man starts off with. First, Leonardo solved it by the Tree Method, getting the answer that the first man has $7\frac{2}{17}$ *denari* and the second $9\frac{14}{17}$.

He then said he would solve it by an alternative approach—what he called the "Direct Method" (*regula recta*): One begins by calling the sought-after quantity a "thing" (*res*) and then forms an equation (expressed in words) that is solved step-by-step to give the answer. Expressed symbolically, this is precisely the modern algebraic method. Without modern symbolic notation, Leonardo's solution looks complicated:

> In solving problems there is a certain method called direct that is used by the Arabs, and the method is a laudable and valuable method, for by it many problems are solved. If you wish to use the method in this problem, then you put it that the second man has the thing and the 7 denari which the first man takes, and you understand that the thing is unknown, and you wish to find it, and because the first man, having the 7 denari, has five times as much as the second man, it follows necessarily that the first man has five things minus 7 denari because he will have 7 of the denari of the second; thus he will have five whole things, and to the second will remain one thing, and this the first will have five times it; therefore if from the first man's portion is added 5 to the second's that he takes, then the second will certainly have 12 denari and the thing, and to the first will remain five things minus 12 denari, and thus the second has sevenfold the first; that is because one thing and 12 denari are sevenfold five things minus 12 denari; therefore five things minus twelve denari are multiplied by the 7, yielding 35 things less 7 solidi* that is equal to one thing and one solidi; therefore if to

* Recall that 12 *denari* made 1 *solidus*.

> both parts are added 7 solidi, then there will be thirty-five things equal to one thing and 8 solidi because if equals are added to equals, then the results will be equal. Again if equals are subtracted from equals, then those which remain will be equal; if from the above written two parts are subtracted one thing, then there will remain 34 things equal to 8 solidi; therefore if you will divide the 8 solidi by the 34, then you will have $2\frac{14}{17}$ denari for each thing; therefore the second has $9\frac{14}{17}$ denari, as he has one thing and 7 denari. Similarly, if from five things, namely from the product of the $2\frac{14}{17}$ by 5, are subtracted 7 denari, then there will remain $7\frac{2}{17}$ denari for the first man, as we found above.[16]

To a modern reader, this does not look much like algebra. But Leonardo's phrases "if equals are added to equals, then the results will be equal" and "if equals are subtracted from equals, then those which remain will be equal" provide a clue to what he was doing. When Leonardo wrote "thing" (in Latin *res* or *cosa*), he meant "the unknown quantity", or more familiarly, x. In essence, Leonardo was formulating two linear equations and then solving them by adding and subtracting equals to each side and by substitution.

Avoiding the introduction of *solidi* (Leonardo replaced 12 *denari* by 1 *solidus*) and working throughout with *denari*, a present-day solution would look like this:

> Let A be the amount the first man has, B the amount of the second. The conditions of the problem can be expressed as the equations

$$A + 7 = 5(B - 7)$$
$$B + 5 = 7(A - 5)$$

These can be rearranged to give

$$A = 5B - 42$$
$$B = 7A - 40$$

Use the second equation to replace B in the first:

$$A = 5(7A - 40) - 42$$
$$A = 35A - 200 - 42$$
$$34A = 242$$
$$A = \frac{121}{17} = 7\tfrac{2}{17}$$

Substituting for A in the earlier equation for B:

$$B = 7\left(\frac{121}{17}\right) - 40 = \frac{847}{17} - \frac{680}{17} = \frac{167}{17} = 9\tfrac{14}{17}$$

This is not exactly a translation of Leonardo's calculation into symbolic form, since he had only one unknown, whereas today we find it more natural to start with two, one for each man in the problem. But this is just a symbolic technicality, though it does raise the interesting question—to which we have no answer—of how he came up with his particular "thing".

Liber abbaci ends very abruptly with the passage:

> And let us say to you, I multiplied 30-fold of a census by 30, and that which resulted was equal to the sum of 30 denari and 30-fold the same census; you put the thing for the census, and you multiply the 30 things by the 30 yielding 900 things that are equal to 30 things plus 30 denari; you take away the 30 things from both parts; there will remain 870 things equal to 30 denari; you

therefore divide the 30 by the 870 yielding $\frac{1}{29}$ denaro for the amount of the thing.

Leonardo was discussing an arithmetic problem solved by algebra. "Census" is the Latin translation of the Arabic *mal* (amount of money, or fund), and is used in many Arabic arithmetic problems, no matter what method is used to solve it. "Mal" (and "census") is also the name of the second degree unknown in medieval algebra.

To a modern reader, this ending may seem a little strange. There is no conclusion, no summary, no reflection on what has been accomplished, no suggestion of new work to be done or of new things to try. Leonardo simply completed his description of the solution to yet another problem and then stopped writing. In fact, many medieval mathematical texts ended in this manner. The most an author might add was a brief statement about the copyist and the date, perhaps with thanks to God. Authors did not write conclusions the way they do today.

But what for Leonardo was the end of a project marked the beginning of an arithmetic revolution that would sweep through Europe. To be sure, one should not overdramatize the effect of just one man and one book. Western Europe, and Italy in particular, had entered a period of significant change before Leonardo was born, in commerce, finance, and education. Many of those changes depended on arithmetic, which provided a highly receptive environment for *Liber abbaci* to have an impact. There is good reason to believe Leonardo was very aware of that situation and wrote his book precisely to meet that need. Certainly, by the time he wrote his second edition—the only one we can read today—he knew of the importance of his work, as is reflected in the way he presented it.

Hindu-Arabic arithmetic was slowly finding its way across

the Mediterranean by conduits other than *Liber abbaci* and would surely have come to dominate European trade, commerce, and finance eventually, even if Leonardo had not written his book.

The greatness of *Liber abbaci* is due to its quality, its comprehensive nature, and its timeliness: It was good, it provided merchants, bankers, businesspeople, and scholars with everything they needed to know about the new arithmetic methods, and it was the first to do so. Though there were to be many smaller, derivative texts that would explain practical arithmetic, it would be almost three hundred years before a book of comparable depth and comprehensiveness would be written (Luca Pacioli's *Summa de arithmetica, geometria, proportioni et proportionalità*, published in Venice in 1494).

CHAPTER 6

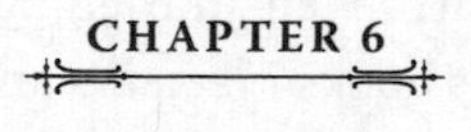

Fame

FOLLOWING THE COMPLETION of the first edition of *Liber abbaci,* Leonardo eventually became something of a celebrity, not just in his hometown of Pisa but throughout Italy. Much of his fame came from the depth of his scholarship. Other books described the Hindu-Arabic number system, a few dating from before Leonardo's time, many more written later, and many of them found their way into classrooms for use as texts. But as the mathematician Laurence Sigler wrote in the introduction to his English-language translation of Leonardo's masterpiece, none were as "comprehensive, theoretical, and excellent as the *Liber abbaci* of Leonardo Pisano."[1]

Between the publication of the first edition of *Liber abbaci* in 1202 and the second edition in 1228, Leonardo published a number of other works that added to his reputation. *De practica geometrie,* (Practical geometry), completed in 1223, was written for professional people whose work involved surveying and land measurement.[2] Though not as long as *Liber abbaci,* it was still a substantial work, and like Leonardo's arithmetic text it included both practical instructions for performing various calculations written for artisans, as well as mathematical verifications of the methods he described for scholars. Leonardo dedicated the book to his friend Dominicus Hispanus, a mathematician in Frederick's court. It contains a large collection of geometry problems, arranged into eight chapters, with theorems based on Euclid's books *Elements* and *On Divisions.* Equations are expressed geometrically, in narrative form, so that $4x - x^2 = 3$ is described as "If from the sum of the four sides, the square surface is subtracted, then three rods remain." Leonardo also included practical information for surveyors, giving the measurements used in Pisa, and a chapter on how to calculate the height of tall objects using similar triangles. In the final chapter he presented what he called "geometrical subtleties", of which he wrote: "Among those included is the calculation of the sides of the pentagon and the decagon from the diameter of circumscribed and inscribed circles; the inverse calculation is also given, as well as that of the sides from the surfaces . . . to complete the section on equilateral triangles, a rectangle and a square are inscribed in such a triangle and their sides are algebraically calculated."

Overall, the book appears to have been heavily influenced by the work of Abraham bar Hiyya, who lived in Spain in the twelfth century. (Bar Hiyya wrote in Hebrew, but his work was one of the many translated into Latin during that period.)

In 1225, Leonardo published a book titled *Flos* (The flower), largely devoted to algebra, and containing his solutions to a series of problems posed to him in a contest organized for Frederick, to whom he sent a copy. *Liber quadratorum* (The book of squares), which he also published in 1225, is a book on advanced algebra and number theory, and is Leonardo's most mathematically impressive work.[3] It deals mainly with the solution of various kinds of equations involving squares, generally with more than one variable, where the solutions have to be whole numbers—the very kind of problem that led Pierre de Fermat to propose his famous "Last Theorem" in the seventeenth century,

The Holy Roman Emperor Frederick II.

Pila citta ſuperbiſsima & potente in Toſcana, gia nimica emula & contraria del populo fi/
rentino, eſsendo queſt'anno oppreſsata da crudeliſsime guerre diuento ſuggetta del preſato
populo firentino, & coſi e gia durata circa centodiecinoue anni ſott'el giugo & gouerno della
inclita ſignoria Firentina. Queſta citta nelli anni paſsati ſu comperata con molta pecunia per

This sixteenth-century drawing of Pisa almost certainly shows the city much as it was in Leonardo's time.

Harbor fortifications at Porto Pisano, the ancient port of Pisa when Leonardo was alive, now part of Livorno. (Keith Devlin)

The Arno River in modern-day Pisa, looking upriver. (Keith Devlin)

The statue of Leonardo in the Camposanto in Pisa today. (Keith Devlin)

One of two historic customhouses in Pisa dating from Leonardo's time. (Keith Devlin)

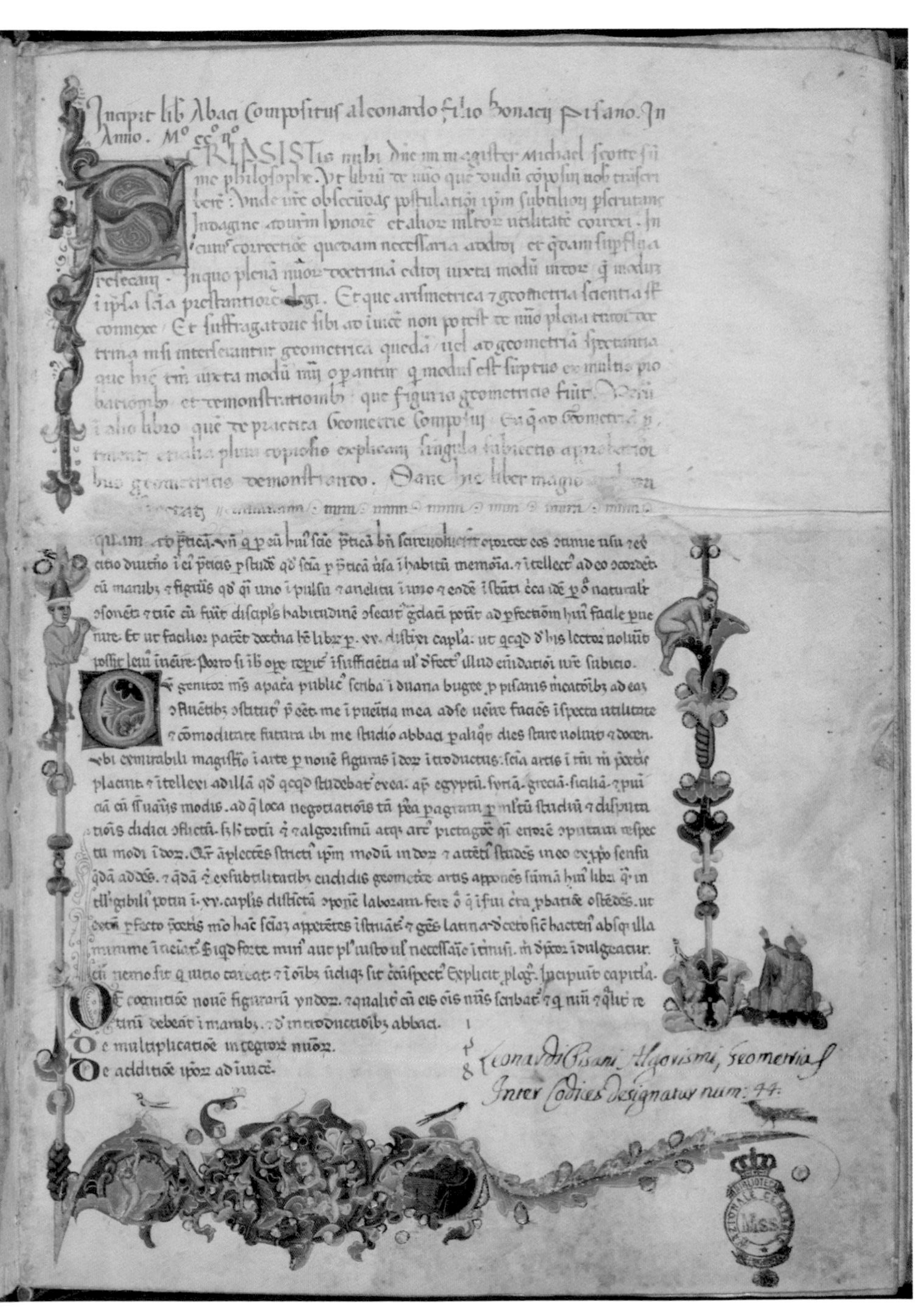
Incipit lib Abaci Compositus a leonardo filio Bonacij pisano. In Anno. M° cc° ii°

Leonardi Pisani Algorismi, Geometria &c
Inter Codices designatus num: 44.

The first page of the Florence manuscript of *Liber abbaci*.
(Courtesy of Biblioteca Nazionale Centrale di Firenze)

A diagram illustrating hand calculation in the Siena manuscript of *Liber abbaci*.
(Courtesy of Biblioteca Comunale degli Intronati, Siena)

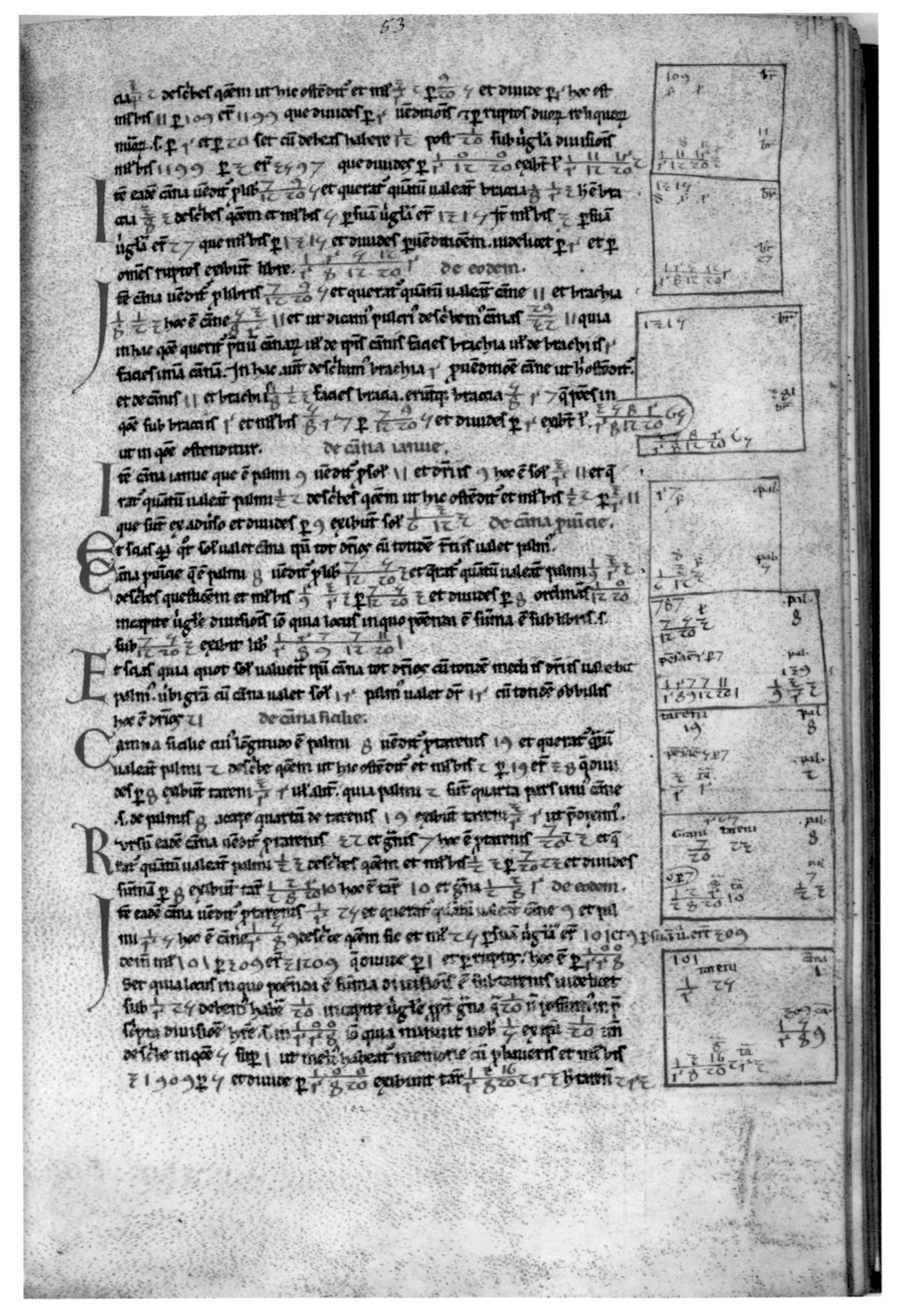

In *Liber abbaci*, Leonardo provided symbolic calculations in the margins to help explain the methods described verbally in the text, almost as illustrations, a practice that was common until symbols gradually became more central in mathematical exposition, as they are today. From the Siena manuscript. (Courtesy of Biblioteca Comunale degli Intronati, Siena)

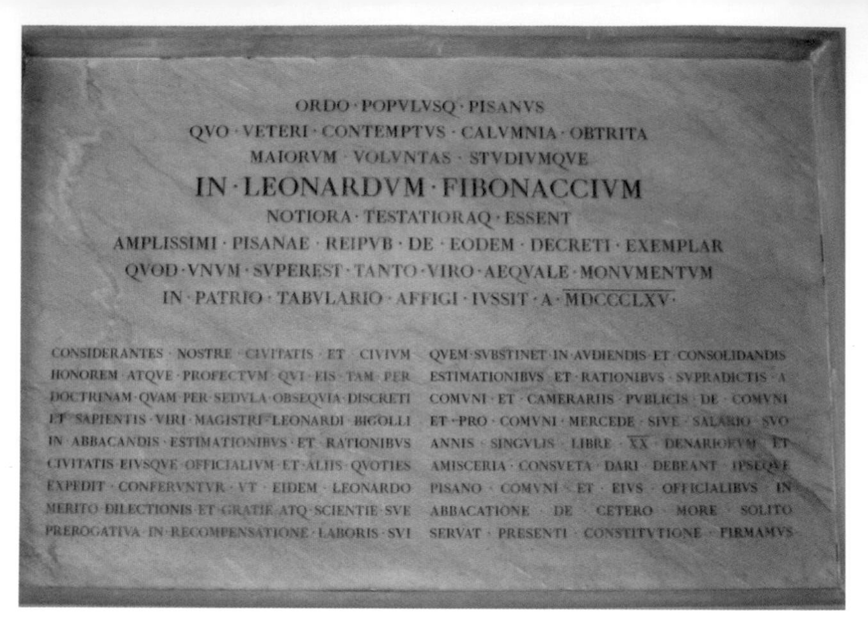

A marble tablet dedicated in 1867 to Leonardo, at the entrance to the State Archives in Pisa. (Keith Devlin)

Ruins of the Central Bridge and the square surrounding the Logge di Bianchi in Pisa following a fierce battle at the end of World War II. Miraculously, the statue of Leonardo (seen in the distance) survived the destruction with only minor damage to its fingers.

a conundrum that was eventually solved by Andrew Wiles in 1994.

These four books, together with a letter he wrote to Theodorus Physicus, the imperial philosopher (*Epistola ad Magistrum Theodorum*),[4] comprise the only surviving works of Leonardo. The letter to Theodorus is undated and is preserved as a copy written in Milan in 1225. In it, Leonardo discussed three problems, one arithmetic, one geometric, and one algebraic. The first of the three is called the "Problem of the 100 Birds," and Leonardo had already presented a solution in *Liber abbaci*. This time, however, he developed a general method for the solution of such indeterminate problems. The second problem is to inscribe a regular pentagon inside an equilateral triangle. He solved it in an algebraic fashion. The final problem is a linear equation with five unknowns, for which he gave a formulaic solution.

Other manuscripts Leonardo is known to have written, but which have been lost, include the *Libro di minor guisa*, a book about commercial arithmetic, and a discussion of book 10 of Euclid's *Elements*, in which he provided a numerical discussion of irrational numbers that Euclid had dealt with geometrically.

SO GREAT DID Leonardo's reputation become in the years following the first publication of *Liber abbaci* that sometime in the mid-1220s, most likely around 1225, he had an audience with the emperor. Frederick heard about *Liber abbaci* and its author from the court scholars, among them the court astrologer Michael Scott, the court philosopher Theodorus Physicus, and the imperial astronomer Dominicus Hispanus, the man who suggested to Frederick that he meet Leonardo when the court was next in Pisa.

The occasion would have drawn a lot of attention. Frederick did everything on a grand scale. Popularly known as *Stupor mundi* (wonder of the world), the young king of Germany, Italy, Sicily, and Burgundy always traveled with a large entourage that included foot soldiers, knights, officials, pages, slaves, dancing girls, jugglers, musicians, and eunuchs, together with his exotic menagerie of lions, leopards, panthers, bears, and apes, all led on chains, hunting dogs, hawks, peacocks, parakeets, ostriches, and a giraffe. A caravan of camels transported his supplies, and Frederick himself would ride at the head of the procession, while at the rear an elephant carried a wooden platform on its back on which were perched trumpeters and archers with crossbows.[5]

Born in Sicily in 1194, Frederick had been crowned emperor by the pope in 1220.* As a young man growing up in Sicily he had developed a passion for learning, particularly science and mathematics, interests that led him to found in 1224 the university in Naples that bears his name: University of Naples Frederick II (Universita degli Studi di Napoli Federico II).

In large part because of its location, Sicily had long been a meeting ground for the Christian and Muslim cultures of Europe and North Africa. The island had four official languages—Latin, Greek, Arabic, and French—as well as the Sicilian spoken by the ordinary people. The Sicilian intelligentsia had acquired from the Muslims an interest in science. In particular, Frederick's grandfather, King Roger of Sicily, was in the habit of summoning to his court learned men from many lands, whereupon he would talk with them at length to discover all they knew, and have records made of all that was said. Roger's successors, William I and William II, arranged for the translation into Latin of the ancient Greek texts on mathematics and astronomy, including the works of Euclid, Aristotle, Ptolemy, and Hero of Alexandria—usually from Arabic editions.

---

* He would remain emperor until his death in 1250.

Frederick spoke six languages: Latin, Sicilian, German, French, Greek, and Arabic. He was an avid patron of the arts and acquired from his family and those around him a deep fascination with all matters scientific and mathematical, including astronomy, optics, geometry, algebra, natural science, and alchemy.

From his grandfather he also learned to adopt a skeptical approach, never accepting new knowledge without adequate evidence. For instance, he pursued his interest in natural history in an experimental fashion. He built incubating ovens to study the development of chick embryos, and he sealed the eyes of vultures to see whether they found their food by sight or by smell. After talking with experts in falconry, he wrote a book, *De arte venandi cum avibus* (On the art of hunting with birds), in which he discussed their classification, habits, migration patterns, and physiology.

The young king sent letters to Muslim rulers, phrasing some of his requests for information as puzzles. He contacted scholars in Egypt, Syria, Iraq, Asia Minor, and Yemen, seeking answers to scientific questions. He wrote to al-Ashraf, the sultan of Damascus, posing problems in mathematics and philosophy, to which the sultan replied giving solutions obtained by an accomplished Egyptian scholar. Scholars from many lands visited his court frequently.

Not surprisingly, then, when Frederick learned of Leonardo's work, he summoned the author of *Liber abbaci* to his Pisan palazzo to discuss the book and give a demonstration of his mathematical ability. In the prologue to *Liber quadratorum* (The book of squares), which Leonardo wrote immediately after his meeting with the emperor, to whom he dedicated it, he wrote: "When I heard recently from a report from Pisa and another from the Imperial Court that your sublime majesty deigned to read the book I composed on numbers, and that it pleased you to listen to several subtleties touching on geometry and numbers . . ."[6] In addition to his audience with the emperor, Leonardo was asked to give a public demonstration of mathematical ability, by responding to three challenge problems put to him in advance by Johannes of Palermo, one of the court mathematicians. Leonardo subsequently presented written accounts of his solutions, two in *Flos*, a copy of which he sent to Frederick, and one in *Liber quadratorum.*[*]

Johannes first asked Leonardo to find a rational number (i.e., a whole number or a fraction) such that, when 5 is added to its square, the result is the square of another rational number, and when 5 is subtracted from its square, the answer is also the square of a rational number. He had almost certainly found this question in an Arabic manuscript. The Arabic scholars seemed to like this kind of number puzzle, and many variations were known. This one was particularly tricky (using the techniques known at the time), and the solution Leonardo found (and subsequently published in *Liber quadratorum*) was both long and ingenious. The final answer was $3 + \frac{1}{4} + \frac{1}{6}$, or $\frac{41}{12}$. For its square increased by 5 gives the square of $4\frac{1}{2}$, and decreased by 5 gives the square of $2\frac{7}{12}$.

Johannes's second problem was another of a kind much loved by Arabic scholars, involving the solution of a cubic equation. In modern symbolic notation, the equation Leonardo was asked to solve is $x^3 + 2x^2 + 10x = 20$. (In fact, this very equation can be found in al-Khayyám's *Algebra*, so Johannes was taking a risk that Leonardo had seen it earlier.) Though the ancient Greek mathematicians knew how to solve quadratic equations—using the same techniques taught to schoolchildren ever since—

* It is possible there were other questions that Leonardo did not feel merited a permanent record. Some scholars have speculated that Frederick organized a mathematical tournament, with Leonardo pitted against other mathematicians, but no evidence exists to support that supposition.

cubic equations are far more challenging.[7] (It would be several centuries before general algebraic methods were developed to solve cubic equations.) Algebraic notation was still many centuries away, so Johannes posed the problem in words:

> Find a number such that if it be raised to the third power, and the result added to twice the same number raised to the second power, and if that result be then increased by ten times the number, the answer is twenty.

Leonardo solved the equation by approximation. There is no record of the exact method he used, but there were a number of techniques available at the time. It is likely his argument went something like this: The first step is to observe that the unknown number must be between 1 and 2. For if it were 1, the answer to the computation would be less than 20, and if it were 2 the answer would exceed 20. So a reasonable initial guess is $x=1.5$. The idea now is to iteratively refine this guess, gradually homing in on the correct answer. The key step is deciding how to improve each approximation.

If x solves the equation $x^3+2x^2+10x=20$, then $x(x^2+2x+10)=20$, and so

$$x=\frac{20}{x^2+2x+10}$$

Hence, if x_n is an approximation of the solution that is too high (respectively, too low), then

$$\frac{20}{x_n^2+2x_n+10}$$

is an approximation that is too low (high). It follows that the average of these two approximations

$$x_{n+1} = \frac{1}{2}\left[x_n + \frac{20}{x_n^2 + 2x_n + 10}\right]$$

is a better one. Thus, starting with an initial approximation of $x_0 = 1.5$ (say), one calculates a sequence of approximations x_0, $x_1, x_2, \ldots, x_n, \ldots$ which fairly quickly reaches an acceptably accurate approximation. Computing to fifteen decimal places of accuracy, this process yields the values:

1.5
1.405737704918030
1.379112302850150
1.371676232676580
1.369605899562000
1.369029978513730
1.368869808289840
1.368825265862720
1.368812879275210
1.368809434619140
1.368808476890990
1.368808210484720
1.368808136235800

and the final approximation in this list turns out to be correct to seven decimal places, after just twelve steps.

Leonardo performed the calculation using sexagesimal fractions (i.e., fractions expressed in base 60), following the practice used by astronomers since Ptolemy. In the sexagesimal notation he was using, the answer he obtained was $1^0 22' 7'' 42''' 33^{IV} 4^{V} 40^{VI}$. In this notation, $22'$ is $22/60$, $7''$ is $7/3600$, $42'''$ is $42/216{,}000$, and so forth, each successive fraction being expressed as a higher power of 60. Expressed in words, as Leonardo presented it to the court, this reads: "One unit, 22 in the first fractional part, 7 in the second, 42 in the third, 33 in the fourth, 4 in the fifth, and 40 in the sixth."

In modern decimal notation, Leonardo's solution is 1.3688081075, which is correct to nine decimal places, a result that is far more accurate than the answer to the same problem that had been obtained (using the same method) by Arab mathematicians who had solved it previously (and more accurate than the answer presented above that was obtained using modern algebra and a computer spreadsheet).

The third problem Leonardo solved was the easiest of the three, being a computation where the unknown quantity is not raised to any power. (In modern parlance, it involves only linear equations.) *Liber abbaci* was full of such problems, though of course Johannes had chosen one that did not appear in Leonardo's own book.

> Three men owned a store of money, their shares being ½, ⅓, and ⅙. But each took some money at random until none was left. Then the first man returned ½ of what he had taken, the second ⅓, the third ⅙. When the money now in the pile was divided equally among the men, each possessed what he was entitled to. How much money was in the original store, and how much did each man take?

Leonardo solved the problem using the Direct Method. He began his solution, as subsequently recorded in *Flos*, by pointing out that

> If you take away half of anything, you have an equal half left; similarly, if you take away a third, that third is

half of the remaining two-thirds; likewise, if you take away a sixth, that sixth is a fifth of the remaining five-sixths.

Let us use the term *res* for the amount each man received when the pile of money was divided equally among them. Then it follows that after the three men had returned the given portions of their money, the first one had half of the money in the original store minus *res*. The second had a third of the original store, minus the same *res*. The third had a sixth of the original store, minus the same *res*.

Since the first man had already put back half of what he originally took, and kept one-half, the half that he kept was equal to one-half the original store minus *res*; in other words, the whole of the money he took from the store was equal to the store minus twice *res*.

Since the second man put back a third of what he had taken, and that third part was half of what he kept [half of the remaining two-thirds], which was a third part of the total store minus *res*, one-half plus one-sixth [or ⅔ reduced to unit fractions] of what he received equaled the third part of said store minus *res*. In other words, the amount the second man took was equal to one-half the total store of money minus one and a half *res*.

Since the third man put back a sixth part of what he took, and that sixth part was a fifth of what he had left [⅙ is a fifth of ⅚] the five-sixths he had left was equal to a sixth part of the total money minus *res*. In other words, the third man took a fifth of the total store minus one and one-fifth *res*.

> Therefore if you add: the total store minus two *res* [the amount the first man took] and half the store minus one and a half *res* [the amount the second man took] and one-fifth the store minus one and one-fifth *res* [the amount the third man took], the total amount, the sum of the amounts the three men took, equals one and seven-tenths of the total money minus four and seven-tenths *res*. Therefore seven-tenths the total store equals four and seven-tenths *res*; therefore, multiplying seven-tenths of the store by ten, and four and seven-tenths *res* by ten, seven times the total store equals forty-seven *res*; therefore if you suppose *res* to equal seven [this amounts to selecting the smallest solution], the total money will be forty-seven [the smallest possible whole-number answer] . . . Therefore since the first man took the total money minus two *res*, or forty-seven minus two *res*, or fourteen, thirty-three will remain for what the first man received. Since the second man took one-half the total money minus one and one-half *res*, twenty-three and one-half, minus ten and one-half, or thirteen, will be what the second man received. Since the third took one-fifth the total money minus one and one-fifth *res*, nine and two-fifths, minus eight and two-fifths, or one, will be what the third man received. And thirty-three of the first man plus thirteen of the second plus one makes forty-seven, the total store.[8]

With that, Leonardo's demonstration was complete. It is hard not to be impressed by Leonardo's ability. Though a mathematician today would regard the three exhibition problems as basic and their solutions as routine, they presented a considerable mental challenge when not approached using modern symbolic notation. A present-day solution to the third problem would translate his words into algebraic symbols:

> Let t be the total money in the store; let u be the amount each man received when the money left in the store was divided equally among them; and let x, y, z be the amounts each man took.
>
> Note first that
>
> $$u = \tfrac{1}{3}(\tfrac{1}{2}x + \tfrac{1}{3}y + \tfrac{1}{6}z)$$
>
> At the end, the first man had what he was entitled to, namely, half the original amount, $\tfrac{1}{2}t$. Therefore, before he received u, he had $\tfrac{1}{2}t - u$. He had already put back half of what he originally took, or $\tfrac{1}{2}x$, and kept $\tfrac{1}{2}x$, so
>
> $$\tfrac{1}{2}x = \tfrac{1}{2}t - u$$
>
> which can be rearranged as
>
> $$x = t - 2u$$
>
> Similarly, the second man ended with $\tfrac{1}{3}t$. Before he was given u, he had $\tfrac{1}{3}t - u$. He had already put back a third of what he originally took, or $\tfrac{1}{3}y$, and kept $\tfrac{2}{3}y$, so
>
> $$\tfrac{2}{3}y = \tfrac{1}{3}t - u$$
>
> which can be rearranged as
>
> $$y = \tfrac{1}{2}t - 1\tfrac{1}{2}u$$
>
> Finally, the third man finished with $\tfrac{1}{6}t$. Before he received u, he had $\tfrac{1}{6}t - u$. He had already returned a sixth of what he first took, or $\tfrac{1}{6}z$, and kept $\tfrac{5}{6}z$, so
>
> $$\tfrac{5}{6}z = \tfrac{1}{6}t - u$$

which can be rearranged as

$$z = \tfrac{1}{5}t - 1\tfrac{1}{5}u.$$

Adding these three equations:

$$\begin{aligned} t &= x + y + z \\ &= t - 2u + \tfrac{1}{2}t - 1\tfrac{1}{2}u + \tfrac{1}{5}t - 1\tfrac{1}{5}u \\ &= 1\tfrac{7}{10}t - 4\tfrac{7}{10}u \end{aligned}$$

Rearranging gives

$$\tfrac{7}{10}t = 4\tfrac{7}{10}u$$

so

$$7t = 47u$$

The smallest whole-number solution to this is where $u=7$ and $t=47$. In this case, the original amount is 47, the first man took $x=t-2u=47-14=33$, the second took $y=\tfrac{1}{2}t-1\tfrac{1}{2}u=13$, and the third took $z=\tfrac{1}{5}t-1\tfrac{1}{5}u=1$. Note that $33+13+1=47$.

Leonardo's triumphant demonstration at the imperial palace is one of the last occasions where we have any reliable knowledge of his activities. The only later reference to the man succeeding generations would refer to as Fibonacci was a proclamation by the commune of Pisa in 1241, granting "the discreet and learned man, Master Leonardo" an annual honorarium of twenty Pisan pounds plus expenses for services to the city.[9] Historians believe that this was in return for Leonardo's service as auditor for the commune.

Frederick died in 1250. Speculation as to how Leonardo ended his days ranges from his being killed in the recurrent civil strife in Pisa to him living out his days peacefully as a revered and honored citizen. In any event, he did not witness

Pisa's glorious period come to an end. In 1284, just forty-three years after the last record of him, Pisa was defeated by its arch rival Genoa in the naval battle of Meloria. Fra Salimbene, a contemporary chronicler from Parma, described the battle with these chilling words: "They grappled their ships together after the fashion of sea fights, and there they fought with such slaughter on either side that even the heavens seemed to weep in compassion, and many on both sides were slain, and many ships sunk. But when the Pisans seemed to have the upper hand, more Genoese came and fell upon them, wearied as they were . . . At last the Pisans, finding themselves worsted, surrendered to the Genoese who slew the wounded and threw the rest into prison." That the captives' lives were spared was not due to any altruism on the part of the Genoese; rather it was to prevent their wives from remarrying back home, thereby reducing the Pisan birth rate.

The Melorian defeat was, however, merely a particularly brutal event that hastened what would in any case have been the end of Pisa's reign as a glorious city-state and a center of maritime trade. Following the death of Frederick, the Hohenstaufen empire, long the source of much of Pisa's security, went into decline, as one by one the emperor's sons and grandsons died at the hands of their enemies. Wars with other cities together with the continuing strife inside the city itself led to a rapid decline in Pisa's trade with the Tuscan interior. On top of that, malaria broke out in the swamps through which Leonardo had ridden to the Porto Pisano when he had embarked on his historic journey to North Africa, decimating the local population.

Meanwhile, Florence, the center of the growing wool industry, rose to become the leading city in Tuscany, while to the north Venice was taking over as the new trading capital of the world. Pisa slid rapidly to the status of the provincial town it has been ever since. But the real glory of Pisa is an intellectual one that lives on through the enormous legacies to humankind left by the city's two most brilliant sons: Galileo Galilei in the sixteenth century, and before him Leonardo Pisano in the thirteenth.

CHAPTER 7

The Fibonacci Aftermath

TOWARD THE END OF THE THIRTEENTH century, and perhaps during Leonardo's lifetime, there appeared the first of what by the fifteenth century may have amounted to a thousand or more arithmetic instructional manuscripts written in vernacular Italian for nonexperts, in particular the commercial community. Some four hundred of them have survived to this day, but any estimate of the total number written can be at best an intelligent guess, since in all likelihood the majority has been lost. In terms of contents, they were all similar. Like *Liber abbaci*, they explained how to write numbers using the ten digits 0 to 9, how place value works, and how to calculate with whole numbers and fractions. They all went on to provide copious worked examples, most of them practical business problems, and, also like *Liber abbaci*, their authors often provided multiplication tables and tables of square roots to facilitate the solution to more complex problems. But they were much shorter than Leonardo's masterpiece, averaging about a hundred manuscript leaves (with text on both sides). A typical leaf would contain from one to five worked problems, giving a total of around four hundred problems for the whole book.

These books became known as *libri d'abbaco* (abbacus books), or *trattati d'abbaco* (abbacus tracts). Most of them did not bear their author's name. They were often illustrated, and some carried annotations saying they were presented as gifts to patrons and important merchants. They were clearly written for a local audience, since the monetary units used in the problems they presented tended to concentrate on the currency of a particular town or region. The author often began by promising to explain "the art of the abbacus as it exists in the town of . . ." In the case of the more poorly written examples, we don't know why they were written or for whom; in some cases they may have been personal notebooks, not intended for use by anyone but the anonymous writer.

The proliferation of abbacus books was remarkable. Yet hardly anyone today knew of their existence until the historian Gino Arrighi started to publish transcriptions of their contents during the 1960s.[1] In 1980, the historian Warren Van Egmond assembled and published a catalog of more than 250 extant Italian abbacus manuscripts published up to 1600.[2] The books in Van Egmond's catalog show an overall growth in this new genre: one in the twenty-five-year period 1276–1300, eight in the next quarter century 1301–25, ten in 1326–50, six in 1351–75, nineteen in 1376–1400, sixteen in 1401–25, thirty-nine in 1426–50, fifty-six in 1451–75, sixty-six in 1476–1500, and the remainder after 1500. Although extrapolating from the manuscripts that survived can provide at best a very approximate estimate of the actual numbers of abbacus manuscripts produced, it is clear that the Italian commercial world had considerable, and growing, interest in learning the new arithmetic.

Whereas many abbacus books did little more than present worked examples—templates to be used by businesspeople without the need for any real learning or understanding—some authors went a bit deeper into the mathematics, producing books that likely were intended for use by teachers in schools as references and sources of examples. Those more didactic books typically began with a short preface explaining the nature and utility of mathematics.

One curious feature of the abbacus books is their continued use of the rhetorical form. *Liber abbaci* had after all shown the Western world how to write numbers in symbolic Hindu-Arabic form and how to carry out computations symbolically. This speaks in part to the educational context in which the abbacus treatises were written, which focused on oral instruction, memorization, and recitation. The goal was to commit facts to memory. Manuscripts were viewed as a means of capturing a spoken lesson. This explains why problems in abbacus texts were often prefaced with a statement such as "If I say to you," and when an intermediate result was to be set aside for later use, the student would be instructed, "So remember it." In their written explanations of computations, the earlier Arabic writers even wrote out numbers in words, rather than use symbolic numerals. When they did give a symbolic presentation, its purpose was more that of a diagram, to show the reader what a person calculating might write on a slate.

A second factor that impeded adoption of symbolic notation was the need for copies of manuscripts to be made by hand. Professional scribes produced elegant-looking copies, but they were not always familiar with the subject. In particular, many did not understand the function of the symbolic representations and would mutilate, abbreviate, or completely omit them.[3] One way for an author to prevent the mathematics being lost was to avoid using symbols. Only with the advent of the printing press in the fifteenth century did authors start to use symbolic expressions.

Although the majority of the problems in the abbacus books were of a business nature, their authors followed Leonardo's example of including a number of more whimsical challenges.* Problems of pursuit were a popular choice. For example:

> A fox is 40 paces ahead of a dog, and 3 paces of the latter are 5 paces of the former. I ask in how many paces the dog will reach the fox.†

Until the late fifteenth century, all the abbacus books were copied by hand, and thus had limited circulation. The first printed abbacus book, an anonymous, untitled volume known today as the *Treviso Arithmetic,* was printed at Treviso, near Venice, in 1478—just twenty-four years after the printing of the Gutenberg Bible in Germany. Written for a general audience by an unknown author in the common Venetian dialect of the period, the *Treviso Arithmetic* used commercial problems to explain the operations of addition, subtraction, multiplication, and division.[4]

A more substantial printed book in the vernacular abbacus tradition was the scholarly work *Summa de arithmetica, geometria, proportioni et proportionalità*, written by the mathematician Luca Pacioli just a few years later, in 1494. A significant difference between Pacioli's book and *Treviso Arithmetic* is that Pacioli dealt with negative numbers. The concept of negative numbers was new in Europe, and Pacioli is believed to have provided the first printed explanation. The unknown author of

* A famous problem of this kind in *Liber abbaci* is about a growing rabbit population. It is discussed in chapter 9 of this book.

† From a fourteenth-century abbacus book by Piero della Francesca, a leading artist of his time, also known as Paolo dell'Abacco. The translation is by Arrighi (1964, p. 78). It is simple to solve using modern symbolic algebra, which reduces the puzzle to the solution of two simultaneous equations, $d = f + 40$ and $5f = 3d$. The answer is 100 fox paces or 60 dog paces.

Treviso Arithmetic, when explaining subtraction, sidestepped the issue by always placing the larger number above the smaller.

Today, mathematicians view Pacioli's book as a landmark in the history of mathematics. Like *Liber abbaci*, it was a mammoth text, filling some six hundred densely printed folio pages. Also like Leonardo's earlier work, it treated arithmetic from both theoretical and practical perspectives. Yet for all its mathematical sophistication, it was clearly designed to be used as a "how to" book by the commercial world. It contained multiplication tables up to 60×60, it provided tables of various monies, it discussed weights and measures used in the different Italian states, and it provided one of the earliest explanations of double-entry bookkeeping. In common with many abbacus books, it also included a "tariff" for the business community. Tariffs, or *libri di mercantia*, were lists of practical information such as the weights and measures and the monetary systems used in the major European and Mediterranean trading cities, itineraries of trade routes, lists of major fairs, and the like. Such lists were also published separately. The first printed tariff was Giorgio di Lorenzo Chiarini's *Libro che tratta di mercanzie et usanze paesi*, published in Florence in 1481; Pacioli included part of it in his *Summa*.

That an able mathematician like Pacioli would write a book on basic arithmetic and practical algebra aimed at the commercial traders underscores the fact that throughout the thirteenth, fourteenth, and fifteenth centuries and beyond, the expansion of mathematics—in particular algebra—in Europe was largely driven by the business world. The authors of the better abbacus books took the algebraic methods Leonardo described in his books and passed them on as a practical tool—which is also how the Arabs had viewed algebra, and why they developed it. Algebra was not taught as an academic discipline in European universities until the middle of the sixteenth century. According to the historian Diana Finiello Zervas, all the mathematics needed to design and build the famous Florentine baptistry door could be found in a typical abbacus book.[5]

PARALLELING THE GROWTH of the abbacus books was the growth throughout Italy of arithmetic schools, called *scuole d'abbaco* or *botteghe d'abbaco* (abbacus schools), where young children were taught how to use the Hindu-Arabic number system.[6] The earliest record of an abbacus school is in the statutes of the commune of Verona in 1294, which mention the appointment of maestro Lotto of Florence to teach mathematics. The teachers in such schools were known as *maestri d'abbaco* (teachers of abbacus, or more colloquially, arithmetic teachers). In addition to teaching children in the schools, many of them taught other adults, who would learn to use the system in the commercial world or would themselves become organizers of, and teachers in, the schools.

Italian boys typically attended elementary school between the ages of six and eleven. At that point, their parents could choose between sending them to a grammar school or an abbacus school. In the former, which typically lasted four to five years, they mastered Latin grammar and read Latin texts, preparing for a career as a cleric, notary, lawyer, physician, or grammar-school teacher, where a solid grounding in Latin was essential. The abbacus schools, which lasted two years, were designed to educate future businessmen by teaching them mathematical and accounting skills. Both texts and teaching were in the vernacular. Some parents sought to get the best of both worlds, by first sending their sons to a grammar school for two to three years, and then to an abbacus school for two more. Leonardo da Vinci and Niccolò Machiavelli were both taught in abbacus schools,[7] da Vinci receiving instruction from Benedetto da

Firenze, who, in addition to being a *maestro d'abbaco*, was the author of a much-copied and much-plagiarized abbacus book, *Trattato d'abacho*.[8]

The *maestri d'abbaco* followed a specified syllabus, typically composed of reading and writing in the vernacular, arithmetic, geometry, bookkeeping, and occasionally navigation. The most detailed syllabus known today is from the school of Cristofano di Gherardo di Dino, who taught in Pisa in 1442:

> This is the way of teaching the abacus in Pisa, from the beginning to the end of the students' learning period, as we will say.
>
> 1. At first, when the boy begins school, he is taught how to make figures, that is 9, 8, 7, 6, 5, 4, 3, 2, 1;
> 2. Then he is taught how to keep numbers in his hands, that is his left hand units and tens and in his right hand hundreds and thousands;
> 3. Then to draw numbers on tables: that is of two figures what it means, and then three figures, four figures and so on. Then how to keep them in one's hand.
> 4. Then one explains the tables of multiplication. One draws it on the table, starting from one times one until ten times ten one hundred, and students learn it very well by heart.
> 5. Then one teaches how to make divisions;
> 6. Then how to multiply fractions;
> 7. Then how to sum fractions;
> 8. Then how to divide [fractions];
> 9. Then how to accrue simple interests and the "new year's merit";
> 10. Then how to measure lands or how to square a number;

11. Then how to make simple discounts and new year's discounts;
12. Then how to calculate the ounces of silver;
13. Then the melting of silver;
14. Then one makes the comparison between the two amounts;
15. And note that to make the above-mentioned calculations, students are to use pencils according to their level. And sometimes have them sum with their hands, or else on the blackboard; occasionally give them some extraordinary homework, according to the teacher's will.
16. Please, note also this general rule: every evening give them homework for the following day according to their level. And, in case of days of rest, homework is to be doubled.

Records indicate that between 1340 and 1510, some twenty or so abbacus schools were operating in Florence alone, with between 1,000 and 1,200 students attending an abbacus school in the one year 1343, a significant proportion of the school-aged male population at the time.[9] A similar situation could be found in Venice, Milan, Pisa, Siena, Lucca, and all the other major cities of Italy. A passage in Giovanni Villani's Florentine *Cronica* suggests that just a few years previously, in 1338, around 10 percent of all boys in the small town of San Giovanni attended the mercantile abbacus school, and 5 percent or so went to the Latin-based grammar school (with more than half of all children learning to read).[10] Though low from today's perspective, these attendance figures indicate a substantial growth in the number of arithmetically skilled individuals for the times.

In the largest cities, such as Venice or Florence, the abbacus schools were owned and operated by private teachers who took in pupils on a daily basis for a fee paid directly by their parents. More typically, abbacus schools were formed by fathers who were wealthy merchants (and thus had influence on public governments) so that their sons could be educated in commercial mathematics. Communal governments then proceeded to attract *abbaco* masters from elsewhere. They drew up contracts that specified the number of years an appointed master had to work, the number of students he was allowed to teach, and the percentage of fees he was to return to the commune. Contracts usually lasted for one to three years. In return, communes would grant masters tax exemptions (full or half), rights to collect fees for tuition, textbooks, and school supplies, and a house to live in. Some abbacus teachers also tutored privately in homes for a fee.

Once students at an abbacus school had learned the basics of the Hindu-Arabic number system and its arithmetic, they were shown how to solve practical problems, such as the everyday exchange of different types of goods or currencies. Other problems might deal with the distribution of profits, where each member invested a certain sum and may have later withdrawn a portion of that amount. Labor contracts were studied, where the employer agreed to a certain wage over the course of a certain term for a certain type of work that produced a specific amount of goods. Lessons on bookkeeping taught students to note weight, length, size, and other quantitative and qualitative information of goods. Almost certainly, most of the abbacus textbooks were intended to be used as references by teachers in those schools. With each problem's solution immediately following its statement, as in *Liber abbaci*, they clearly were not designed to be used as textbooks, where the student does the problems and then checks the answer with a solution at the back of the book, as with a present-day textbook.

THE ABBACUS BOOKS and the abbacus schools grew to meet a demand caused by changes in Italian commercial life at the time—changes that required a more numerate citizenry. When Leonardo was growing up, commerce was still carried out by individual traders who took their goods to the markets. This required only the simplest kind of arithmetic, easily performed using finger reckoning or with a traditional counter abacus. During the thirteenth century, however, a radical new way of doing business developed. Sea travel became faster and more reliable with improved navigation—due largely to the introduction of the magnetic compass and the development of the portolan chart, a navigational map based on realistic descriptions of coasts and harbors. (The word "portolan" comes from the Italian adjective *portolano*, meaning "related to ports or harbors".) Sea transportation also grew safer with the suppression of piracy. The beginning of what we would today regard as a modern, international banking system provided financial instruments such as letters or credit and bills of exchange, and the newly developing marine insurance industry provided protection against loss of a vessel in a storm. The development of modern double-entry bookkeeping by bankers in Florence facilitated the growth of more complex forms of business organization.

All of this led to the formation of trading conglomerates, headed by executive managers who traded not from the deck of a ship or in a foreign marketplace but from their homes in major Italian commercial centers such as Pisa, Florence, and Venice. While others handled the actual goods, this new kind of trader dealt with account books, letters, and bills of exchange. That new world required symbolic arithmetic, as described by Leonardo in *Liber abbaci* and taught by the *maestri d'abbaco* in the abbacus schools. The old methods using Roman numerals and the counter abacus simply were not up to the task.

Though the superiority of Hindu-Arabic numerals and arithmetic led to their rapid adoption by many traders, for some time public opposition prevented them from being universally accepted. The reason typically given was that the new numerals made merchants' books difficult to read. Some bookkeepers objected on the grounds that the numerals were vulnerable to being altered. According to one Venetian manual on bookkeeping, "The old figures alone are used because they cannot be falsified as easily as those of the new art of computation, of which one can, with ease, make one out of another, such as turning the zero into a 6 or a 9, and similarly many others can also be falsified."[11] Italian courts gave legal precedence to documents bearing Roman numerals over those written in the Hindu-Arabic system.

One group with a clear vested interest in preventing change was the select few trained in the use of a mechanical abacus, and some of them did try to hold the new methods at bay. For example, in the statutes of the Arte del Cambio (Guild of Money Changers) of Florence in 1299, a proclamation forbade its members from using the new numerals.[12] And in 1348, the University of Padua insisted that lists of its books have their prices given in Roman numerals—"not by figures, but by clear letters," the order stipulated.[13]

As a result of the opposition, Roman numerals continued to be used in Italian merchants' ledgers for some time. There were occasional uses of mixed forms too, such as $\mathrm{II}^{m}\mathrm{III}^{c}\mathrm{XV}$ for 2315, a combination of place value and Roman numerals. But change was inevitable. The Hindu-Arabic system was simply a far more efficient way to handle numbers, and by and large most Italian merchants had made the switch by the late 1400s.

The driving force behind the adoption of more efficient ways to do arithmetic was the growth of the Italian commercial world. From the eleventh century to the fifteenth, the per capita income of northern Italy tripled. By the end of the fourteenth century, many Italian commercial enterprises had global reach, with some 150 Italian banking firms operating multinationally. Of course, arithmetic was not the cause of this growth, nor was it the only ingredient. Its role was more like that of oxygen in making a fire. It takes fuel to start and maintain a large blaze, and it is easy to overlook the invisible gas that is crucially involved. But without sufficient oxygen, all you have are slow-burning embers.

WITH PISA IN decline not long after Leonardo's death, by the fifteenth century Florence had become a major locus for much of the activity in the new arithmetic and its application, particularly in the field of finance. Insight into the growth in use of the Hindu-Arabic numbers in finance comes from examining the ledgers of the Medici Bank.[14] The Medici family, mentioned for the first time in a document of 1230, came from the agricultural Mugello region, north of Florence. Their rise to prominence began under Cosimo de' Medici during the late fourteenth century. Having acquired their wealth in the textile trade, in 1397 they founded the Medici Bank, which became Europe's largest and most respected bank during the fifteenth century, when it had at least eight trading houses throughout Europe. The Medicis' lasting contribution to business, commerce, and accounting was the development of the double-entry bookkeeping system for tracking credits and debits. Although its origins can be traced back to Roman times, the system in its modern form was first used by accountants working for the Medici family in Florence. In the Medici account books from 1406 onward, Hindu-Arabic numbers began to appear frequently in the narrative or descriptive column. From 1439 onward they replaced

Roman numerals in primary entries (journals, wastebooks, etc.), but not until 1482 were Roman numerals abandoned in the final business ledger of all but one Medici merchant. From 1494 onward, only Hindu-Arabic numerals were used in all Medici account books.[15]

The Hindu-Arabic system took longer to migrate beyond Italy's borders, however. In 1494, the money changers in Frankfurt attempted to prohibit its use just as the Florentines had two centuries earlier, persuading the city to issue an ordinance that proclaimed "the master calculators are to abstain from calculating with digits."[16] Abacus-board arithmetic was still dominant in northern Europe up to the end of the sixteenth century. But in time, the sheer superiority of the new system came to dominate there just as it had in Italy. The driving force behind the change was reflected in the popular term for the Hindu-Arabic numerals in Italy: *figura mercantesco* (mercantile figures)—a force that clearly transcended national boundaries. Just as had occurred with Leonardo in Bugia, the northern Europeans came to recognize the power of the new arithmetic through business dealings. Many of those interactions took place in what was by then the commercial capital not just of Italy but the entire world: Venice.

Even in Leonardo's time, the city of canals threatened to surpass Pisa as the most important commercial center in Europe. As one commentor wrote of the city in 1267, "Merchandise passes through this noble city as water flows through fountains."[17] By the fifteenth century, Venice, now a republic, had risen to be the undisputed trade capital of the world. It ruled over a large mainland empire that included, in addition to the city of Venice itself, Ravenna, Treviso, Padua, Vicenza, Verona, and Brescia. The Venetian merchants did business with most of Europe as far north as the North Sea and the Baltic, including the major international centers of Ghent, Bruges, Antwerp, Amsterdam, and London, as well as the Arabic-speaking world to the south and the east across the Mediterranean.

Along with its status as the world's commercial capital, from the fourteenth century onward, Venice was viewed as the well-spring of new mathematically based methods for doing business, and merchants came from all over Europe to learn the *arta dela mercandanta*, the Italians' mercantile art. In Germany, the initial proponents of the new arithmetic were called "cossists", after the Latin word *cosa* for the thing or unknown of algebra. Soon, German businessmen came in increasing numbers to study the *Welsche Praktik*, the foreign practices of business, commercial arithmetic, and currency exchange. In fact, so many German merchants came to Venice to do business that the city built a special *fondaco* to accommodate them, the Fondaco dei Tedeschi (German *fondaco*), a five-story structure with a large courtyard that provided accommodation for more than eighty visiting merchants and their servants at a time. They came to learn about all the new business practices the Italians were developing and using to great effect in the areas of banking, finance, insurance, exchange, wholesale trade, manufacturing, colonial administration. The commercial world was rapidly turning into the global marketplace we today take for granted.

Underlying it all was the new arithmetic that Leonardo had imported from Bugia. The fact that the first printed abbacus book (the *Treviso Arithmetic*) was produced in the vicinity of Venice was no accident. (Treviso is a small town just to the west of Venice that the Republic had acquired in 1339.) Venice was where the demand for this powerful new tool was by far the greatest.[18] As the unknown author of the *Treviso Arithmetic* wrote at the very end of his book, "By as much study as you have given the work, by so much has it appealed to your ardent desires, I do not doubt it will bring back to you much fruit."[19]

By the end of the sixteenth century, all of Europe used the Hindu-Arabic system, and as a result began its inexorable rise to world domination in trade and finance. This was the start of the commercial revolution—a revolution that depended on powerful, efficient ways to handle numbers. The arithmetic revolution itself was over. But who should get the credit for starting it? It is tempting to say Leonardo—the timing was right, *Liber abbaci* was accurate and comprehensive, providing many practical examples, and Leonardo and his books achieved fame in his lifetime. But was that the case?

CHAPTER 8

Whose Revolution?

FOLLOWING PIETRO COSSALI's discovery of Leonardo's work at the end of the eighteenth century, many scholars speculated that the growth of the abbacus books and the abbacus schools—and hence the arithmetic revolution they generated—were direct consequences of the publication of *Liber abbaci*. In *Summa de arithmetica, geometria, proportioni et proportionalità*, the book in which Cossali first found reference to Leonardo, the author, Luca Pacioli, had credited the Pisan for instigating the European arithmetic revolution, a view Cossali endorsed in his book. If what they suggested were true, Leonardo's role in history would be on a par with that of Copernicus or Galileo or Kepler.

Certainly, the contents and structure of *Liber abbaci* made it an excellent source for the material that drove the revolution. And the timing of events, particularly the growth pattern revealed by Van Egmond's quarter-century counts of abbacus books, makes Pacioli's conclusion plausible. Yet despite the fact that medieval authors copied freely from one another all the time, neither *Summa* nor the majority of the abbacus books contained any passages from *Liber abbaci*. To credit the revolution to Leonardo, further evidence would be required. In particular, scholars would have to identify the path by which the contents of *Liber abbaci* were transmitted to the pages of the early abbacus books. Assembling that evidence took historians and archivists many years of painstaking effort, along with some remarkable strokes of luck. The final piece of the puzzle fell into place as recently as 2003.

The easiest case to make is that, whether or not Leonardo was the instigator of the arithmetic revolution, he was at least its inspiration. By the time he completed the second edition of *Liber abbaci*, Leonardo was famous throughout Italy as a great mathematician and an honored citizen of Pisa. As a result, many people were aware of his book and doubtless wanted to learn of its contents—merchants, scholars, teachers, and parents, as well as the emperor Frederick. In short, the very fact that *Liber abbaci* had been written by someone of Leonardo's stature, coupled with a general description of its contents and an appreciation that what he taught was useful in commerce, could well have created a demand for more easily digestible explanations.

When individuals of far lesser mathematical talents than Leonardo sought to meet that demand with books of their own, for the most part they looked elsewhere than in the dense pages of *Liber abbaci*. In some instances, no doubt, this was because the budding abbacus authors found *Liber abbaci* far too challenging; in other cases the problem was that they could not read Latin; and some probably sought other sources because they simply could not get hold of one of the few available copies of Leonardo's manuscript. When authors of abbacus books referred to Leonardo as their inspiration, as many did,* they

* This practice had come to an end by the time printed texts were produced, and Pacioli was the only author of a printed abbacus book who gave such credit.

may have been simply using his name to add an air of authority to their work, but the very fact that they did so shows that Leonardo was regarded as the primary figure in the tradition.

Even the Danish historian Jens Høyrup—one of the few scholars who have seriously questioned Leonardo's role as the original source for the abbacus books—finds this scenario plausible. When the famous sixteenth-century mathematician Girolamo Cardano (one of the early pioneers of what we now call probability theory) wrote his great work *Ars magna* in 1545, he credited Leonardo for introducing modern arithmetic and algebra into Europe. That led Høyrup to speculate, in a 2005 paper: "Instead of being the starting point of *abbaco* culture Fibonacci may have been an extraordinary representative who, growing, had grown taller and more conspicuous than any other representative—so tall that Cardano saw nobody but him in the landscape who was worth mentioning."[1]

By the time Høyrup's paper had been published, however, a scholar in Italy had already identified the crucial missing text connecting *Liber abbaci* to the abbacus books—thereby proving that Leonardo was not just the inspiration for the arithmetic revolution but also the instigator in terms of its content and form. To find that missing link, the Italian scholar had to delve deeply into the highly specialized world of archival forensics—the detailed comparison of texts to uncover the historical flow of ideas. Working for many years in the Italian archives, she assembled sufficient evidence not only to identify, by name and author, a key manuscript—long since lost—that led from *Liber abbaci* to the abbacus books, but also to say with considerable certainty what its contents were. What makes this particular investigation of added interest to the layperson today is that the scholar was uncovering the origins of the modern world we live in. A typical abbacus book looks remarkably like the schoolbooks we all learned from. Replace the old-style language and the local cultural references with those of today, and (apart from the full-color illustrations that today's publishers are so enamored by) you find yourself looking at a typical modern arithmetic textbook. As Warren Van Egmond wrote in his 1980 catalog of abbacus books that made the scholastic world aware of the abbacus tradition, "However, the tradition does not end [in the sixteenth century] . . . the abbaci are only the beginning of a tradition of arithmetical writing that extends to the present day. The elementary books that we all used to learn basic arithmetic are the direct descendants of the abbacus books of the fourteenth and fifteenth centuries."[2]

ONE OBSTACLE THE mathematical historian faces stems from the way mathematics is recorded and transmitted. To the mathematician, for much of the time, doing mathematics seems to be (some would prefer to say "is") a process not of invention but of discovery. When the mathematical community views an advance as "routine," the individual who made it is quickly forgotten. The path was there, waiting to be discovered, and someone had to find it first, but exactly who did so was largely a matter of happenstance and not worthy of further mention. The important thing is the path, which others can now follow. The only exception is when someone establishes a new formula or proves a particular theorem, but only if finding the formula or proving the theorem is deemed to be important and to have required considerable ingenuity.

This is very different from many other areas of human creativity, where we rightly acknowledge the individuals who do things first. In many cases, the creative act is clearly unique. If Shakespeare had not lived, for example, *Hamlet* would never have been written. In contrast, if Euclid had not proved that there are infinitely many primes, someone else would have. The only uncertainty is how long it would have taken. (As it happened, we do cite Euclid for that theorem, but only because it is an important result and the proof showed great originality and ingenuity—though it's not entirely clear that Euclid himself did discover the result.)

Hindu-Arabic arithmetic falls into the category of something waiting to be found. The individuals who found it first are, on the whole, not remembered, nor are they acknowledged by others who came later and built on their work. Use of the term "Hindu-Arabic arithmetic" reflects the fact that the Indians developed the system and the Arabs refined it, but with a few exceptions there is little record of exactly when and by whom various advances were made.

Thus, reading an ancient manuscript written by a mathematician is unlikely, on its own, to provide a historian with information about the sources the mathematician consulted in preparing it. For instance, Leonardo clearly obtained much of his material for *Liber abbaci* from the works of al-Khwārizmī, either directly or indirectly, yet he made no mention of that fact apart from a brief remark in the final chapter. Commentators outside the field of mathematics find this strange, and sometimes seek an explanation. Yet Leonardo was simply following a mathematical practice as old as the subject itself.

If that lack of proper accreditation were not problem enough for the historian, often when a name does become attached to a particular mathematical advance it turns out to be the wrong name. Pythagoras's theorem is a good illustration; the result was known long before Pythagoras of Samos was born.

One way medievalists trace the development of ideas is by making a detailed comparison of the manuscripts, looking for clues such as passages that one author copied directly from another. Medieval authors frequently copied entire passages from an existing manuscript without crediting the source. If a medievalist finds identical passages in two texts, he or she can conclude that either one author copied from the other or both copied from a third. (This last possibility requires further investigation if the original manuscript is lost, leaving only the two that contain material copied from it.)

For many centuries, copying of books was done almost exclusively by monks in the more scholarly monasteries. It was a slow process. It could take a year or more to make one copy of a book the length of the Bible. From the thirteenth century onward, with the rise of the universities creating a much larger demand for books, bookmaking gradually moved out of the cloisters and into the commercial world. Professional copy companies appeared, usually on the edges of the new universities—an early forerunner of today's commercial photocopy shops. Some of them grew quite large; one early fifteenth-century copy shop in Florence employed forty-five lay copyists.

Hand copying produced some attractive-looking manuscripts. With considerable time to devote to the task, the scribes, who in Leonardo's day worked mainly for spiritual reward, developed elaborate calligraphic styles and would often let their creativity run free when it came to adorning and illuminating the page with colorful swirls, drawings, and other embellishments to the text. When a scribe completed a book, he often closed with a personal note. One particularly memorable example—though not in *Liber abbaci*—reads:

Explicit hoc totum;
Pro Christo da mihi potum,

which can be translated as:

The job is done, I think;
For Christ's sake give me a drink.[3]

Unfortunately, the scribes' creativity frequently extended to the book's contents as well. They were not averse to making changes, leaving parts out, or copying material from another manuscript, generally without any indication that they had done so. And of course, there was always the possibility they would simply make a copying mistake. All of which makes present-day medieval scholarship fiendishly difficult.

The more influential texts were, of course, copied many times, so errors could—and did—multiply, as copies were made of copies. Faced with a collection of manuscripts all clearly versions of the same original text, the medievalist has the task of putting them in correct sequence. Even if a manuscript gives a date, there can be uncertainty whether it refers to the completion of the original text or when the copy was made. In fact, most of the abbacus texts bear neither the name of the author nor a date, but the authors generally phrased their examples in terms of the local coinage and units of weight—which varied from location to location and over time—so with some painstaking detective work it has been possible to assign a location and date to many of the manuscripts.

Writing history by comparing manuscripts has an obvious limitation: It can be applied only to manuscripts that either have survived or are referred to substantively in a work that has survived. It is always possible that a key manuscript was lost. Sometimes, with a lot of painstaking effort and perhaps a bit of luck, it is possible to infer a particular manuscript's existence and even determine its author and contents. This is what happened with the birth of the abbacus texts.

If the abbacus manuscripts had contained material taken directly from *Liber abbaci*, the lineage would have been clear from the outset. To be sure, a few abbacus books were essentially just vernacular extracts from *Liber abbaci*; but they came later, after many other abbacus texts were in circulation.[4] The vast majority of abbacus books, including all the known early ones, had almost nothing in common with Leonardo's masterpiece. In particular, they typically lacked the organization, precision, and expository standard of Leonardo's treatise and were much less extensive. Clearly, the abbacus authors must have obtained their material elsewhere. In which case, the first task facing the historian was to decide whether the genre began with material taken from one work or several sources.

In the introduction to his 1980 survey, Van Egmond noted that there was little duplication from one abbacus book to another—the abbacus authors were not simply copying one another. That seems to point to multiple initial sources, but appearances can be deceptive. With roughly four hundred problems in each book, taken altogether the abbacus books provided around 150,000 seemingly different worked examples, but when you ignore the exact words and numbers used and focus on the underlying arithmetic problems, you find a fairly small number. Moreover, the presentations of the problems all followed a standard pattern that varied little over the three-hundred-year extent of the abbacus books. A typical worked problem began with a brief description of a situation, followed by a question, and then the solution. For example:

> A *solidus* of Provins is worth 40 *denari* of Pisa and a *solidus imperiali* is worth 32 of Pisa. Tell me how much will I have of these two monies mixed together for 200 *lire* of Pisa? Do it thus: add together 40 and 32 making 72 [*denari*] which are 6 *solidi* and divide 200 *lire* by 6 which gives 33 *lire* and 6 *solidi* and 8 *denari*, and you will have this much of each of these two monies, that is 33 *lire* 6 *solidi* 8 *denari* for the said 200 *lire* of Pisa. And it has been done.[5]

The author would always introduce the question by a phrase such as "Dimmi" (Tell me), "Domandoti" (I ask you), or "Voglio sapere" (I want to know). In many of the earlier manuscripts, the entire problem statement would be preceded by an introductory phrase like "Fammi questa regola" (Do this problem for me) or, if the problem is used to illustrate a rule that has just been introduced, "Voti dare essempro alla detta regola" (I want to give you an example of this rule). The solution to the problem usually began with a phrase such as "Fa cosi" (Do it thus) or "Dei cosi fare" (You must do it this way). The author would often include a restatement of the answer after it had been obtained. The ending too was standard, with a phrase such as "Ed e fatta" (And it has been done) or "Et cosi fa tutte le simigliante ragione" (And do all similar problems in this way).

Thus, when scholars subjected the abbacus books to a closer look, what had initially seemed like evidence of multiple sources turned out to be highly suggestive of at most a small number of original texts, with the differences between the abbacus books being largely in depth, sophistication, and accuracy, and localization to a particular town or region. There were, however, still several possibilities for the original texts.

During the twelfth and thirteenth centuries, a wave of Arabic science flowed into Europe, via Latin translations made in Spain, Italy, and the Crusader states. For instance, in Montpellier, many scribes were translating and copying texts, among them al-Khwārizmī's *Arithmetic*. One early book in Latin that some historians have suggested was an abbacus source is *De algorismo*, written by the English scholar John of Halifax. Born around 1195, in England (presumably in the northern town Halifax), John was educated at the University of Oxford, after which he went to live in Paris, becoming a professor of mathematics at the university. Known also as Johannes de Sacrobosco, he wrote a number of books on mathematics and astronomy. *De algorismo* was an elementary exposition (in Latin) of Hindu-Arabic arithmetic and included chapters on addition, subtraction, multiplication, division, square roots, and cube roots. Though some abbacus authors may have consulted it, however, there is no firm evidence it was the initial source for the majority of abbacus books.

Since later abbacus books may well have been based on earlier ones, historians trying to identify the origins of the arithmetic revolution focused much of their attention on the earliest manuscripts in the genre. The first vernacular abbacus book may have been a *libro di nuovi conti* (book of new calculations), written around 1260 in Siena, but no copy has survived.[6] The oldest abbacus books still in existence date from around 1290. One of them is *Livero de l'abbecho.*[7] Its unknown author described it as "lo livero del abbecho secondo la oppenione de maestro Leonardo dela chasa degl'figluogle Bonacie da Pisa" (abbacus book according to the opinion of master Leonardo Fibonacci), perhaps making it one of the earliest to make such use of Leonardo's name. Another is *Columbia algorism,*[8] which was transcribed by Kurt Vogel in 1977 as *Ein italianishes Rechenbuch aud dem 14. Jahrhundert.*[9] The German phrase "14. Jahrhundert" in Vogel's title translates as "14th century", and that underscores the difficulties facing the medieval historian. At the time Vogel published his translation, he believed it was a later manuscript, but subsequent examination of the coins it mentioned showed that it was almost certainly written earlier, most likely around 1290.[10] Paolo Gherardi's book *Libro di ragioni* (Book of problems), written in Montpellier in 1328, is the oldest known vernacular Italian abbacus book with a chapter on algebra.

In all cases, however, while examination of early abbacus manuscripts suggested some that may have been used as sources for others that were written later, none indicated an original source other than Leonardo, and over the years the majority of scholars came to view the abbacus books and the abbacus schools as part of a tradition initiated by the appearance of *Liber abbaci.* For instance, in the introduction to his 1980 catalog of abbacus books, Van Egmond declared, "All the manuscripts and books contained in the present catalog can be regarded as members of this tradition [abbacus books] and direct descendants of Leonardo's book."[11] In his analysis, he wrote: "The printed abbaci are by and large a direct continuation of a tradition that began with the *Liber abbaci* in 1202. The abbacus books thus represent a continuous and remarkably uniform tradition that stretches from the work of Leonardo Pisano to the end of the sixteenth century, shifting from handwritten to printed books as the primary means of publication changed from manuscript to printing."[12]

In a similar vein, Kurt Vogel, in a highly regarded, authoritative article on Leonardo observed:

> In surveying Leonardo's activity, one sees him decisively take the role of a pioneer in the revival of mathematics in the Christian West. Like no one before him he gave fresh consideration to the ancient knowledge and independently furthered it. In arithmetic he showed superior ability in computations. Moreover, he offered material to his readers in a systematic way and ordered his examples from the easier to the more difficult . . .
>
> With Leonardo a new epoch in Western mathematics began . . . Leonardo became the teacher of the masters of computation (the *maestri d'abbaco*) and of the surveyors, as one learns from the *Summa* of Luca Pacioli, who often refers to Leonardo. These two chief works were copied from the fourteenth to the sixteenth centuries.[13]

A similar conclusion was also expressed in no uncertain terms by the mathematical historian Ivor Grattan-Guinness: "The immediate origin of this tradition lies unquestionably in the Latin *Liber abbaci* ('Book of the Abbacus') written in 1202 by Leonardo of Pisa—more familiar to modern readers by his nickname, Fibonacci (Boncompagni 1857)."[14]

Finally, Laurence Sigler, the mathematician who translated *Liber abbaci* into English, says in his introduction:

> For three centuries or so a curriculum based upon Leonardo's *Liber abaci* was taught in Tuscany in schools of abaco normally attended by boys intending to be merchants or by others desiring to learn mathematics. Other instructors and some very good mathematicians also wrote books of abaco for use in the school. These books vary from primitive rule manuals up to mathematics books of quality, but none was so comprehensive, theoretical, and excellent as the *Liber abaci* of Leonardo Pisano.[15]

These modern scholars were presumably influenced by the few abbacus books that do have sections resembling *Liber abbaci*; for example, two mid-fifteenth-century Florentine treatises bearing the common title *Praticha d'arismetricha,* one written around 1450 by an anonymous Florentine master who describes himself as a pupil of Domenico d'Aghostino, the other written about ten years later by Benedetto of Florence.* Both treatises provided exhaustive summaries of the type of mathematics that was being taught in the abbacus schools, in the

---

* The first is included in codex Palatino 573 of the Biblioteca Nazionale of Florence; the second is part of manuscript L.IV.21 of the Biblioteca Comunale of Siena.

same way as *Liber abbaci* described the type of mathematics being developed in the Arabian countries that Leonardo had visited. Both authors demonstrate their thorough knowledge of Leonardo's works, which are often cited and from which they quote large excerpts. In the introduction, the anonymous author of the 1450 work places Leonardo first in the list of authors "to be considered" and invokes the Pisan's authority in many passages of the work. The fifth part of the treatise, "where we will study cases of enjoyment," is taken entirely out of chapter 12 of *Liber abbaci* (as the author states). Also the eighth part, "which contains the calculation of roots," includes vernacular translations of passages in chapter 14 of *Liber abbaci*. The final section of the treatise, devoted to algebra, presents "cases written by Lionardo Pisano, perfect arithmetician," and covers problems in the third part of chapter 15 of *Liber abbaci*. This section opens with the following declaration:

> Lionardo Pisano, as he is known for a text in the great volume titled *Praticha d'arismetrica*, studied in Egypt and there perfected his knowledge by investigating the subject. And in these Tuscan regions, he was the first to bring knowledge and to announce the rules. And this is announced through the words of Master Antonio in the book of his "Fioretti," where he demonstrates that the intellect of said Lionardo from Pisa is great. Lionardo wrote many books for our science, among which were the ones I know, namely: the *Libro di merchaanti detto di minor guisa*, the *Libro de' fiori*, the *Libro de' numeri quadrati*, the *Libro sopra il 10° d'Euclide*, the *Libro di Praticha di geometria*, the *Libro di Praticha d'arismetricha* from which I took what I want to write about presently. And these works are in Santo Spirito and in Santa Maria Novella, and also in the Abbey of

Florence, and they are highly regarded by many of our citizens.[16]

Clearly, with these two manuscripts, we see a link to *Liber abbaci*, but both were published 250 years after Leonardo's text, so neither can have played a role in initiating the abbacus movement. Though most scholars believed that all the abbacus books were descendants of *Liber abbaci*, no one could show how the transmission had taken place for the texts that did not resemble Leonardo's masterpiece—the vast majority. A crucial first step was missing: Someone—and the simplest explanation was a single individual—must have taken material presented in *Liber abbaci* in a highly sophisticated way, and reformulated it in the much simpler fashion found in the abbacus books. That key missing text, if it existed, would have been the very first abbacus book, and the true spark that ignited the arithmetic revolution.

A CRUCIAL CLUE came from looking not for similarities between *Liber abbaci* and the abbacus books but at a major difference. A feature of many abbacus books that is not found in *Liber abbaci* is the inclusion of a section on geometry, where methods are described for finding lengths or areas or volumes of geometric figures, with examples typically being cast in terms of calculating the height of a tower by triangulation, the area of a field, or the volume of a pond or a barrel. What scholars found particularly curious about this departure from *Liber abbaci* is that it was the only one. When they started to ask themselves how it could have come about, they found themselves on a path that would eventually lead to the missing link.

Where did the abbacus authors get the material on geometry? Leonardo had touched on some geometric issues in chapter

15 of *Liber abbaci* in order to use geometric methods to solve problems in arithmetic and algebra, but left a more comprehensive treatment of geometry to another book, *De practica geometrie*, completed in 1220. Comparing the geometry in abbacus books with that in *De practica geometrie* revealed the same similarities and differences as those between the arithmetic in the abbacus books and that in *Liber abbaci*. In other words, the first abbacus book—if there was a single source—was almost certainly a simplified abridgement of not one of Leonardo's books (*Liber abbaci*) but two, *Liber abbaci* and *De practica geometrie*. The historian Elisabetta Ulivi summarized the evidence in 2002: "[The abbacus books] were written in the vernaculars of the various regions, often in Tuscan vernacular, taking as their models the two important works of Leonardo Pisano, the *Liber abaci* and the *Practica geometrie*."[17] If someone wrote an initial, introductory-level summary of Leonardo's two masterworks that began the abbacus tradition, that person would have to have been not only proficient in Latin but also sufficiently skilled in mathematics to read and understand the contents of two long, dense, and highly sophisticated scholarly texts, select some of their contents, and recast the material in a much simpler fashion.

The most obvious candidate was Leonardo himself, and it was known that he did write a simplified version of *Liber abbaci*. On a number of occasions, he referred to having written a *Liber minoris guise* (Book in a smaller manner). One such reference is in *Liber abbaci* itself,[18] another is in his *Liber quadratorum*, and a third in his *Flos*. A further reference to the same lost work occurs in the fifteenth-century abbacus book *Praticha d'arismetricha*,[19] whose anonymous author referred to it as *Libro di minor guisa o libro di merchaanti* (Book in a smaller manner or book for merchants). That description of it as a "book for merchants" is significant, since it suggests that the *Libro di minor guisa* would most likely have comprised material from the first ten chapters of *Liber abbaci* together with parts of *De practica geometrie.* But without a surviving copy of Leonardo's "book in a smaller manner", the evidence, while highly suggestive, was not conclusive.

Then, in 2003, the Italian scholar Rafaella Franci published the results of a remarkable study of a manuscript she came across in the Biblioteca Riccardiana in Florence. The manuscript is anonymous and occupies pages 1 to 178 of the library's codex 2404. The work itself is undated, but dates in some of the problems place its writing at around 1290, and the vernacular language used places it in the Umbria region.[20] It may be the earliest vernacular manuscript that has survived to this day. Its author began with the declaration: "This is the book of abacus according to the opinion of master Leonardo of the house of sons of Bonaçie from Pisa." As always, the author's reference to Leonardo cannot be taken as evidence that he used any of the Pisan's writings as sources, though in this case, with the book written not long after Leonardo was alive, the pronouncement should be given some credence.

The book (known as *Livero de l'abbecho* [Book of abbacus] from the anonymous author's introduction) is divided into thirty-one short chapters:

1. Three things
2. Things which are sold by the hundreds
3. The pepper rules
4. Rules with no name
5. Merchants' rules of exchange
6. Rules of barter of money and currency
7. Rules of "marcho Tresçe"
8. How many "cantara" and "charrubbe" and "grane" are in one uncia

9. Buying "bolçone" with currency and in pounds
10. Rules to "consolare" and alloy coins
11. Various rules belonging to "consolare"
12. Rules of interest or usury
13. Rules belonging to the rule of usury
14. Rules of "saldare ragione"
15. Rules of the companies
16. On buying horses
17. On men passing along between one another
18. On men who find purses
19. On men who picked up currency together
20. On rule of "procachio" or travel
21. On men who went earning money at the markets
22. On a cup and its bottom
23. On trees and wooden "vogle"
24. On vases
25. On men who walk together
26. On men who brought pearls for sale in Constantinople
27. On vats and barrels in which the wine comes out of an opening in the bottom
28. On a man who sent his son to Alexandria
29. On a worker who worked on a product
30. On men who walk one after the other
31. On the rules for many tough and light "guise" of many "contintione."

It is not an original work. Roughly three-quarters of the problems are faithful translations into the vernacular of problems in chapters 8, 9, 10, and 11 of *Liber abbaci*. Of particular note to a modern reader, on leaf c.107v, Fibonacci's famous rabbit problem appears, though recast in terms of pigeons. The book was clearly written for a wide audience, since the author began each problem type with simpler problems than those in *Liber abbaci*. Overall, the treatment is both shorter and less complete than that in *Liber abbaci*. The author omitted all of the introductory material that occupies the entire first part of Leonardo's masterpiece, dealing with the Hindu–Arabic representation of numbers, the algorithms for operations, and the methods for calculating with fractions, and proceeded as if these topics were well known to the reader. He began instead by describing the Rule of Three, which he stated thus:

> If there is given any calculation, and if that calculation implies three things, we should multiply the thing we want to know by the thing that is not similar and divide by the other.

Franci pointed out two features important to her analysis. First, the author provided four examples to show how the rule works, the first where the known numbers are all integers, the others involving one, two, and three mixed numbers, respectively. Second, the margins contained tables that summarized the operations to be done, but the diagrams were not explained in the text.

Leonardo's treatment of the Rule of Three in *Liber abbaci* was very different; he stated it in the most general manner where there are four proportional numbers, three of which are known and one of which needs to be determined; he emphasized the homogeneity of the units of measurement of similar items; and, at the end of the treatise, he provided a rationale for the rule by the theory of proportion. The unknown Umbrian did none of these. Moreover, Leonardo introduced a diagram to proceed more quickly in the calculations, analogous to the tables in the margins of *Livero de l'abbecho*, but unlike the Umbrian he went on to explain in great detail the way in which it was created and employed. Moreover, the examples in *Liber abbaci* are different from and more complex than the ones in *Livero de l'abbecho.*

The topics presented by the Umbrian author in the next three chapters—"things which are sold by the hundreds," "the pepper rules," and "the fabrics that are sold by the *channa* and arm"—are also present in *Liber abbaci,* but again are discussed in a simpler fashion, using different examples. Also, whereas Leonardo dealt with problems of merchandise pricing, doubtless with an eye on the international wholesale trade, the Umbrian wrote for a readership more interested in local and retail trading. The same is true of the longer, fifth chapter, devoted to the "rules of exchange"; while *Liber abbaci* had monetary examples from many places, the Umbrian author referred only to the coinage of central and northern Italy, with the exception of one problem referring to coins from Paris.

Up to this point in Franci's analysis, it would have been possible—though highly unlikely—for the anonymous Umbrian to have taken his material from *Liber abbaci.* But the most intriguing aspect of his book—the inclusion of three chapters (12, 13, and 14) on calculating interest and depreciation—makes clear that *Liber abbaci* could not have been the Umbrian's source. In *Liber abbaci,* there is hardly any mention of this topic. In fact, in chapter 14 of *Livero de l'abbecho,* "Rules of 'saldare ragione' ", the author introduced a topic completely absent from Leonardo's treatise, dealing with questions about mercantile practice, including how to handle payments made on different dates. (The problem he gave makes reference to the years 1288, 1289, and 1290, which is how the work can be dated.) This material was obviously included to make the treatise more useful to merchant readers, but from where did he take it?

Throughout the book, the author gives no indication of any particular mathematical skill, nor of the mathematical knowledge that would be required to make intelligent selections of what to include and what not. He must have simply copied the entire book from another work, with at most minor changes. That other work, Franci suggests, could be none other than Leonardo's now lost *Libro di merchaanti detto di minor guisa* (Smaller book for merchants). Part of the reason why such a conclusion can be made with such confidence stems from the nature of mathematics. There is no evidence of anyone living at that time other than Leonardo who had the mathematical ability to write an original work of that kind. With such a demand for arithmetic knowledge, anyone capable of writing a good text would have quickly found an audience and become known.

Taking Franci's supposition to be correct—and by this stage in her investigation the case she had assembled was as good as one can expect in medieval literary forensics, though more evidence was to come—*Livero de l'abbecho* provides a close copy of the *Libro di minor guisa,*[*] thereby establishing Leonardo's lost work as the original source of the hundreds of abbacus books that followed.

Franci's conclusion has now been supported by examinations of other texts written in the first half of the fourteenth century. Despite the fact that they make no meaningful references to Leonardo, they exhibit striking similarities with the Umbrian's treatise, especially when it comes to the explanation of the Rule of Three.

Leonardo may have initiated the abbacus book tradition not only by writing his *Libro di minor guisa,* Franci suggests, but also by lecturing on abbacus methods in Pisa. It is possible, she says, that the first vernacular abbacus book was *Libro di nuovi conti* (Book of new calculations), written around 1260 in Siena by Ugo Ugurgeri, a Cistercian monk from San Galgano, but no

* The book is often referred to by this shorter title that omits the term *di merchaanti.*

copy has survived.[21] Since Siena and Pisa are very close, its author could have had access to a copy of *Libro di minor guisa*. This supposition gains further support from examinations of the contents of abbacus manuscripts written in Pisa,* despite the fact that the earliest known date from the late 1300s, over a century after Leonardo's death.[22]

For example, in one interesting Pisan text found in manuscript 2186 of the Biblioteca Riccardiana of Florence, the author introduced himself as (c.9v):

> I Cristofano di Gherardo di Dino, Pisan citizen of the St. Bastiano chapel in the Chinsica quarter of Pisa, today, May first 1442, with God's name and His salvation began to write this Book of abbacus.
>
> This is the manner and the way to teach the abbacus in the style of Pisa, that is the beginning, middle and end as we will explain later.

Though Cristofano never mentioned Leonardo by name, the book appears to be a vernacular presentation of some parts of Leonardo's work. Other manuscripts too indicate that Leonardo's influence in Pisa was still strong two centuries after his death, with Pisan authors adopting a common way of explaining the abbacus methods. Franci concluded her seminal article: "From our analysis, it appears that, despite the small number of citations of Leonardo Fibonacci in abbacus treatises, the influence of his works—particularly *Liber abbaci* and *Libro di merchaanti*—is absolutely clear both in the contents and in the form of the texts, which follow the two typologies that he introduced: one more didactic and the other more scientific."[23]

---

* Some of the more significant ones are described in the final section of this chapter.

Taken all together, the evidence is overwhelming. Leonardo left two important legacies: One, comprising his scholarly books *Liber abbaci*, *De practica geometrie*, and *Liber quadratorum*, would lead to the development of modern mathematics. The other, his *Libro di minor guisa*, provided the template for all the abbacus books and the associated growth of practical, commercial arithmetic. Leonardo of Pisa started the modern arithmetic revolution.

0 1 2 3 4 5 6 7 8 9

Futher Evidence from the Pisan Manuscripts

Perhaps the most interesting of the Pisan manuscripts that support Franci's conclusions is the anonymous *Tractato dell'arismetica*.[24] The dates 1270–71 and 1315–16 appear in some calculations, which, together with the Pisan coinage and measures referred to, suggests that it was written there between the end of the thirteenth and the beginning of the fourteenth century. Unlike the *Livero de l'abbecho*, this later manuscript begins, like *Liber abbaci*, with an explanation of the representation of numbers in the positional system, use of the hands to do calculations, and a description of algorithms for the four basic arithmetic operations. With those basics out of the way, the topics covered are: Rule of Three Things, ratios of "merito", 6 "chaselle" of currency exchange, the 13 "chaselle", ratios of "regare a termine", discount rules, payments from one town to another, buying from different places, currency exchange, different ways of bartering, rules of the companies, rules of "consolare monete", "merito" on New Year's, ratios of the greater thing, how to calculate the root of numbers, and grouping numbers. The presentation of the decimal numeral system and the representation of numbers with the hands are very detailed and accurate. The algorithms taught for the multiplication and division are those presented by Leonardo in *Liber abbaci*, which the anonymous Pisan author calls *moltiplicare in croce* and *partire a danda*, respectively. They are presented through several examples, accompanied by detailed explanations.

The Rule of Three is stated in a sophisticated fashion (c.8r):

> The rule for doing all the calculations in which there are three things, either currency or weight or measure, is to multiply the thing we want to know by the one that is not the same and divide by the other. One must also keep in mind that two of those three things have to be similar either by name or by substance or, if they are not, they must be adapted, and if they cannot be adapted, then that calculation cannot be done.

This explanation is similar to the Umbrian author's, but the later Pisan stresses the homogeneity of the measures (as did Leonardo in *Liber abbaci*). Overall the Pisan's treatise is more complete and better organized than that of the earlier Umbrian, but the many commonalities suggest the same source, providing further evidence that both authors took their material from Leonardo's *Libro di minor guisa*. A significant difference between the two texts is the inclusion in the Pisan's manuscript of a chapter on algebra, which *Livero de l'abbecho* does not cover. The author's treatment of algebra differs from that presented in *Liber abbaci*, which leads to the supposition that Leonardo included in *Libro di minor guisa* a treatment of algebra that was simpler and more similar to that which subsequently appeared in *Trattato dell'arismetica*. Certainly, all the presentations of algebra that are found in the abbacus books are similar to the one by the anonymous Pisan.

CHAPTER 9

Fibonacci's Legacy—in Stone, Parchment, and Rabbits

THOUGH FORGOTTEN FOR SEVERAL centuries, Leonardo is honored in Pisa today by an imposing stone tablet and a handsome white-marble statue. But his name—or rather his nineteenth-century nickname—is best remembered by virtue of one of the problems he included in *Liber abbaci*. Though it was little more than a recreational puzzle having no great significance, the sequence of numbers that constitute its solution—the Fibonacci numbers—is an important part of Leonardo's cultural legacy.

Toward the latter part of *Liber abbaci*'s chapter 12, nestled between problems involving the division of food and money, Leonardo inserted a whimsical problem about a growing rabbit population. He did not invent the problem; it dates back at least to the Indian mathematicians in the early centuries of the Current Era who developed the number system *Liber abbaci* described. He clearly realized, however, as did his Hindu predecessors, that it is an excellent problem for practicing how to use the new number system. Although later generations of mathematical historians would consider *Liber abbaci* one of the most influential books of all time, to most people its author's greatest fame would rest on this one problem. In what was to become his most famous passage, Leonardo wrote his way into twentieth-century popular culture with these words:

HOW MANY PAIRS OF RABBITS ARE CREATED BY ONE PAIR IN ONE YEAR.

> A certain man had one pair of rabbits together in a certain enclosed place, and one wishes to know how many are created from the pair in one year when it is the nature of them in a single month to bear another pair, and in the second month those born to bear also.[1]

Leonardo wanted the reader to assume that once a pair of rabbits becomes fertile, they continue to produce offspring every month. As usual, he explained the solution in full detail, but the modern reader can rapidly discern the solution method by glancing at the table Leonardo also presented, giving the rabbit population each month:

beginning	1
first	2
second	3
third	5
fourth	8
fifth	13
sixth	21
seventh	34
eighth	55
ninth	89
tenth	144
eleventh	233
twelfth	377

The general rule is that each successive number is the result of adding together the previous two; thus, 1+2=3, 2+3=5, 3+5=8, et cetera. As Leonardo observed at the end of his solution, although he calculated the population at the end of one year (377), this simple rule gives you the population after any number of months.

The numbers generated by the addition process Leonardo described to solve the rabbit problem are known today as the Fibonacci numbers. They first appeared, it seems, in the *Chandahshastra* (The art of prosody) written by the Sanskrit grammarian Pingala sometime between 450 and 200 BCE. Prosody, the rhythm, stress, and intonation of speech, was important in ancient Indian ritual. In the sixth century, the Indian mathematician Virahanka showed how the sequence arises in the analysis of meters with long and short syllables. Subsequently, the Jain philosopher Hemachandra (ca. 1150) composed a text on them.

The Fibonacci numbers were given their name by the French mathematician Edouard Lucas in the 1870s, after his compatriot, the historian Guillaume Libri, gave Leonardo the nickname Fibonacci in 1838. A lot of the fascination with them is due to the surprising frequency with which they arise in nature.

For example, the number of petals on flowers is a Fibonacci number more often than would be expected from pure chance: an iris has 3 petals; primroses, buttercups, wild roses, larkspur, and columbine have 5; delphiniums have 8; ragwort, corn marigold, and cineria 13; asters, black-eyed Susan, and chicory 21; daisies 13, 21, or 34; and Michaelmas daisies 55 or 89.

Sunflower heads, and the bases of pinecones, exhibit spirals going in opposite directions. The sunflower has 21, 34, 55, 89, or 144 clockwise, paired respectively with 34, 55, 89, 144, or

233 counterclockwise; a pinecone has 8 clockwise spirals and 13 counterclockwise. All Fibonacci numbers.

A third example arises in phyllotaxis, which studies the arrangement of leaves on plant stems. As they go up the stem, they spiral round. Start at one leaf and let p be the number of complete turns of the spiral before you find a second leaf directly above the first. Let q be the number of leaves you encounter going from that first one to the last in the process (excluding the first one). The ratio p/q is called the divergence of the plant. Common divergences are: elm, linden, lime, and some common grasses $1/2$; beech, hazel, blackberry, sedges, and some grasses $1/3$; oak, cherry, apple, holly, plum, and common groundsel $2/5$; poplar, rose, pear, and willow $3/8$; almonds, pussy willow, and leeks $5/13$, all ratios of Fibonacci numbers.

The key mathematical fact underlying nature's seeming preference for Fibonacci numbers is their close connection to an equally famous mathematical constant known as the Golden Ratio. Often denoted by the Greek letter Φ (phi), the Golden Ratio is, like the mathematical constant π, an irrational number—a number whose decimal expansion continues forever, without ever settling into a regular, repeating pattern. The decimal expansion of π begins 3.14159; Φ starts out 1.61803.

The number Φ first appeared in Euclid's *Elements* (written around 350 BCE), where it is defined as the ratio into which you should divide a line so the ratio of the entire line to the longer division equals that of the longer division to the shorter. Euclid gave it the name "extreme and mean ratio". In the fifteenth century, the Italian mathematician Luca Pacioli gave it the more evocative name "Divine Proportion", publishing a three-volume work by that title. It acquired the alternative name "Golden Ratio" in 1835, in a book written by the mathematician Martin Ohm.

With two suggestive names, one hinting at God, the other at

wealth, various false beliefs have attached themselves to the number: that it is the aspect ratio of the rectangle the human eye finds the most aesthetically pleasing, leading the ancient Greeks to incorporate it into much of their architecture, including the Parthenon in Athens; that various painters and musicians used it in their compositions; that if you measure the distance from the tip of your head to the floor and divide by the distance from your navel to the floor you get the Golden Ratio; and several more. Some investment companies even claim to achieve better-than-the-rest performance by basing their actions on the Fibonacci sequence. None of these claims holds up under even basic scrutiny.

However, the Golden Ratio is genuinely exhibited by the growth of plants. Nature's inevitable preference for efficiency leads it to place petals on flowers, seeds in flowerheads, and leaves on plant stems in a fashion that depends on the Golden Ratio, which has a mathematical property that results in optimized structure.

The connection between the Fibonacci numbers and the Golden Ratio was first verified in the nineteenth century: If each Fibonacci number is divided by the one that precedes it, the answers you get grow steadily closer to the Golden Ratio—in mathematical terms, the limit of those ratios is the Golden Ratio. (The first few values work out: $2/1 = 2$; $3/2 = 1.5$; $5/3 = 1.666$; $8/5 = 1.6$; $13/8 = 1.625$; $21/13 = 1.615$; $34/21 = 1.619$; $55/34 = 1.618$.) Since Φ is an irrational number, whereas the number of petals, spirals, or stamens in any plant or flower has to be a whole number, Nature "rounds off" to the nearest whole number, and because of the above limit property, this will tend to be a Fibonacci number.

Fascination with Fibonacci numbers led to the formation in 1963 of the Fibonacci Association, which continues to publish a regular mathematical journal, the *Fibonacci Quarterly*. The association and its journal are devoted to mathematical investigations of the Fibonacci numbers and sequences like it. For the most part, this is not deep mathematics; nonetheless, number sequences generated in a fashion like the original Fibonacci sequence turn out to have many fascinating mathematical properties.

IN 1241, THE Comune of Pisa decreed that an amount of money should be given annually to Leonardo for his service to the city. (This proclamation is the only indication that he was still living in 1241.) The text of the proclamation was reproduced on a stone tablet the city erected on June 16, 1867, to honor their great ancestor. It can be found on a wall in the entrance to the Archivio di Stato (State Archives) of Pisa at 30, Lung'arno Mediceo (the road that follows the northern bank of the Arno). The medieval text follows an introductory declaration written in 1865.

The building itself is of interest. Formerly known as the Palazzo Toscanelli, it originally belonged to the Lanfranchi family, but then passed on to Alessandro della Gherardesca, who made a number of architectural modifications. In 1821–22, it was the home of Lord Byron and his circle of friends, during the period in which Pisa was called the "Paradise of Exiles". From there Byron departed for Greece, where he met his death.

The inscription is written in a very formal nineteenth-century form of Latin, which translates literally as:

> The Rulers and People of Pisa in the year 1865 after ignoring old crushing falsehoods and where the will of the Elders was to study what was better known and proven about Leonardo Fibonacci ordered the city archives to file a copy of the decree by the same Most

> Eminent Republic of Pisa that one monument equal to so great a man survive.
>
> [The 1241 decree]
>
> In consideration of the honor brought to the city and its citizens and their betterment by the teaching and zealous cooperation of that discreet and learned man, Master Leonardo Bigolli, as well as by his regular patriotic efforts in civic and patriotic affairs, the Pisan Commune and its Officials in certain right and conscious of our prerogative to make recompense for work that he performed in heeding and consolidating the efforts and affairs already mentioned confer upon this same Leonardo so meritorious of our love and appreciation an annual salary or reward from the Commune of 20 free denari and the usual accompaniments. This we affirm with the present statement.[2]

Once again there appears that name Leonardo Bigolli, this time in a context that certainly would not have meant "blockhead".

BESIDES A FAIRLY recent engraving showing the head and shoulders of a young man, the only image we have of Leonardo is a nineteenth-century statue in Pisa. Both the engraving and the statue are believed to be artistic creations, not based on fact. (The statue, though showing a man in middle age, bears sufficient resemblance to the engraving that the sculptor, Giovanni Paganucci, may have referred to it in his work. The statue, in white marble, is located in the Camposanto (Monumental Cemetery) in the Piazza dei Miracoli near the Leaning Tower. Leonardo's face is sculpted with classic features, an angular face with a strong narrow nose, a well-defined jaw, and a cleft in his chin. He stands erect, draped in a long, cascading tunic, with his head covered by a hood, giving him a vaguely clerical look. Two locks of carefully curled hair peek out from under his hood. He looks down from his plinth with a kind, scholarly expression, holding a book in one hand.* His other hand is stretched forward as if gesturing to emphasize a point to a student. Both hands have fingers missing, due to damage in the Second World War.

The statue stands on a pedestal bearing the inscription:

A Leonardo Fibonacci Insigne Matematico
Pisano del Secolo XII

which translates as "To Leonardo Fibonacci, noted mathematician of Pisa of the 12th Century." (The date is a bit misleading. Although Leonardo was born around 1170, all of his mathematical publications were in the thirteenth century.)

Low down at the rear of the right side of the pedestal is a small inscription that reads:

G. Paganucci
Firenze 1863

The statue has an interesting history.[3] The initiative for creating it was taken not in Pisa but in Florence, by two politicians from ancient aristocratic families in Tuscany: Baron Bettino Ricasoli and Marquis Cosimo Ridolfi. In 1859 the Grand Duke of Tuscany was exiled, and the following year the region was annexed to the kingdom of Savoy, which soon became the new

* Though people did bind manuscripts into "book form" in Leonardo's time, they did not resemble modern books, like the one in the statue's hand.

unified Italian state. During the transition, Tuscany was ruled by a provisional government. Ricasoli was the prime minister, and Ridolfi was the secretary for education. They both were active in promoting culture, including the founding of a modern institute for advanced studies that later became the University of Florence. On September 23, 1859, Ricasoli signed a decree resolving that the state of Tuscany should finance the carving of three statues, for Pisa, Lucca, and Siena. Each was to commemorate an important local person; the statue for Pisa was to be of Fibonacci. The decree cited him as "the initiator of algebraic studies in Europe." The work was commissioned to Giovanni Paganucci, a sculptor in Florence, who completed the Leonardo statue in 1863. It was placed in the Camposanto, alongside the statues of many other prominent Pisans, and officially unveiled on June 17.

The statue remained in the Camposanto until 1926, when the then fascist authorities in Pisa removed it and two other statues and placed them in three squares of the town, where they would serve as a testimonial to some of the great citizens that the city had produced. All three statues relocated were of individuals referred to as "Pisano": Leonardo and a famous father-and-son pair of sculptors and architects, Nicola and Giovanni Pisano. A sentence was added to the inscription on the pedestal of each statue that read:

Dall'oblio alla gloria per volontà fascista

Taking the translation of the Italian word *per* as "for", this translates literally as "From oblivion to glory for fascist will," though a better translation would be "From oblivion to glory thanks to fascist will."

The fascist authorities placed Leonardo in a prominent position, in the front of the Logge di Bianchi, an elegant square at the southern end of the Ponte di Mezzo (Central Bridge), the bridge that crosses the Arno River in the center of the town. A bust of the minor nineteenth-century politician Felice Cavallotti was removed to make room for the square's new occupant, and the square itself was renamed Piazza XX Settembre (20th September Square)—the name commemorating the date in 1870 when the city of Rome, until then under the rule of the popes, was conquered by the new Italian state that had been founded in 1861.

At the end of the Second World War, American and German troops fought an intense, monthlong battle from opposite sides of the Arno, during which the bridge was destroyed, the piazza was badly damaged, and several surrounding buildings were either destroyed or left in partial ruins. But somehow, Leonardo's statue miraculously survived with just minor damage to his hands. A remarkable photograph taken shortly after the battle shows Leonardo standing almost totally unscathed amid a sea of rubble.

When the war was over, the statue of Leonardo was removed to allow for the rebuilding of the bridge and the square. Since the Camposanto itself was partly ruined and under reconstruction, it was not possible to return the statue to its original home. Instead, it was placed inside a city warehouse, where it was left forgotten for two decades.

In 1966, the statue was finally taken out of storage and put on display in the Giardino Scotto,[4] a small public park on the bank of the Arno, adjacent to a stretch of road named Lung'arno Leonardo Fibonacci. Unfortunately, the Giardino is a popular visiting spot not just for people but for birds, and over the years the statue became not only badly discolored by the riverside weather but also covered in bird droppings. A photograph of the statue in the park, dark and dirty, which can be found on many Web sites, was taken by Frank Johnson at that time.

Around 1990, the statue was taken away and restored, and eventually placed back in the Camposanto where it began.

A FAR MORE significant part of Leonardo's legacy than the Fibonacci sequence, the memorial tablet, or the statue is the corpus of fourteen medieval *Liber abbaci* manuscripts that have survived. Since the originals of Leonardo's 1202 and 1228 editions are lost, these fourteen extant copies of Leonardo's 1228 edition are the closest we can come to his original masterpiece.

Seven are mere fragments, consisting of between one and a half and three of the book's fifteen chapters, and four others are somewhat fragmented. Of these, six are housed in Florence, three in Paris, and one each in Milan and Naples.* The remaining three are complete or almost so and are all in Italy. One, in the BNCF in Florence,[5] is believed to date from the late thirteenth or perhaps the early fourteenth century. The manuscript is complete, which probably explains why the publisher Baldassarre Boncompagni used it as the basis for his first printed edition in the mid-nineteenth century, even though it is perhaps not the oldest. Another, from which chapter 10 is missing, is housed in the Vatican Library in Rome[6] and is believed to date back to the late thirteenth century. The third is in the Biblioteca Comunale di Siena (Siena Public Library)[7] and is generally believed to date from the thirteenth century, possibly 1275, and according to some scholars may be the oldest—though others have suggested it may have been written as much as a century later.

* The locations are the Biblioteca Nazionale Centrale di Firenze (BNCF—Florence National Central Library, four copies), the Biblioteca Laurenziana Gadd in Florence, the Biblioteca Riccardiana in Florence, the Biblioteca Ambrosiana in Milan, the Biblioteca Nazionale Centrale in Naples, the Bibliothèque Mazarine in Paris, and the Bibliothèque Nationale de France in Paris (two copies). For more details, see Hughes 2004.

Perhaps surprisingly, none of the early manuscripts is in Pisa, but there is a plausible explanation. In the days when every copy of a manuscript had to be prepared laboriously by hand, there would have been little incentive to keep a copy in a city that already had a manuscript, Leonardo's own originals. Thus any copies made would most likely have been for use elsewhere. Rome, Florence, and Siena are obvious cities where copies would be found.

During the course of my research into Leonardo, I visited Italy a number of times, and on two occasions I arranged to examine the manuscripts in Siena and Florence. (The Vatican Library has been closed for extensive renovation for some years, and it has not been possible to gain access to the rare manuscripts.) I knew the contents, having read Sigler's English translation. But I wanted to know what it felt like to hold in my hands a manuscript of the work that had fascinated me for so long, a manuscript that, while not in the author's own hand, was copied close to his lifetime—the scribe may have started work while Leonardo was still living—and would thus look and feel similar to the original.

The Biblioteca Comunale, where the Siena manuscript is kept, is at number 5 Via della Sapienza (Street of Knowledge), a few minutes' walk from the Piazza del Campo that is the heart of the city. The manuscript there is large and heavy, measuring over 20.5 centimeters wide by 30 centimeters high by 5 centimeters thick, discounting the heavy cover which is a discolored brown bearing gold lettering. The spine bears the inscription

LION. PISANI
DE ABACO

near the top, and the Siena Public Library reference number

L.IV.20

near the bottom. From the pages of clean, thick white paper that form the inner binding, it is clear that the volume has been rebound fairly recently.

The manuscript, missing chapter 15, comprises 224 sheets, each one written on both sides. The page numbers—running from 1 to 224, appropriately in Hindu-Arabic numerals—were added later, center-top of the front side of each sheet. The parchment is thick and stiff but not brittle. Apart from the front page, which has partly disintegrated and is attached to a backing sheet, each page is in remarkably good condition. An occasional page has a hole in it, and the outside edges of some pages have worn away under the cumulative influence of eight hundred years' worth of page-turning hands. Many pages are a whitish cream color, others a light brown, and many are discolored. Each page had been carefully ruled with a grid to guide the lettering, and the text is written in brown ink, with every numeral in red. At the start of some paragraphs, the initial letter is enlarged and also in red. In addition, the scribe embellished some chapter openings with large stylized letters in blue and gold.

Despite the damage, the first page still shows the original title near the top:

Aritmetica Leonardi Bigholli de Pisa

The final page of the manuscript ends with the legend

Exptir cbet aritmetica Leonardo bigholli de pisa

squeezed in at the bottom of the page.

The Florence manuscript is housed in the Biblioteca Nazionale Centrale di Firenze, which is on a small square on the north bank of the Arno River, in the center of Florence. This manuscript is surprisingly thinner than the one in Siena, although about the same height and width. Its cover bears no inscription and fastens with two brass clips. Each page is 20 centimeters wide by 30 centimeters high, the same size as the one in Siena. Discounting the cover, it measures 4 centimeters thick, a full centimeter less than the copy in Siena. There are 213 pages, each written on both sides, giving 426 sides altogether, almost the same number as the Siena copy, even though the Florence manuscript includes chapter 15.

On the inside of the front binding are pasted two tiny fragments of manuscript, all that is left of the original cover page. All I could make out, and then not with total certainty, was:

Leonardi Pisani Algorism A[ritm]*etica*

The first page of the manuscript bears the following legend at the top of the page:

A C Leonardus pisanus Algorisma & Geometrie est
Abbacie florenty

References to the Florence manuscript generally refer to it as "badly faded," but apart from some pages, including the first, this is definitely not what I found. On the contrary, overall the manuscript seemed to be in much better condition than the one in Siena. The paper is dry but not brittle, and felt slightly thinner than that of the Siena copy—which explains why the bound volume is much thinner, even though the page count is almost the same. None of the pages have holes or worn edges, as they do in Siena. The Latin text is in a brown-black ink, and all numerals are in red. The scribe who had written it certainly had the greater artistic flair of the two, decorating the margins of many pages with fancy swirls, and making much more extensive use of elaborate red letters to begin paragraphs and large red, blue, and ornate gold letters to start new sections. Some of the pages are faded, and many have turned brown or gray and become spotty. On some pages you can see the guidelines the scribe had used to line up the text. But overall it is in great shape, given its age.

It's a strange feeling to hold a book written by a medieval scribe over seven hundred years ago, particularly one that played such a major part in the development of the modern world. I had been familiar with the contents for many years and was aware of the circumstances that led to the writing of the manuscript, but that was cognitive and intellectual. Holding such a manuscript in my hands, as I did in Siena and Florence, was physical and visceral—knowledge through the body as opposed to knowledge in the mind. On each occasion, as I slowly turned the pages, Leonardo and his story became real.

Every age produces a few individuals who are both very much ahead of their time and also of their time—the former to imagine what is possible, the latter to make it happen. Figures such as Archimedes, Copernicus, Galileo, Kepler, Newton, and Einstein. Leonardo Pisano, Fibonacci, deserves to be in their midst. History almost forgot him, only a nickname given to him by a later historian surviving, and then only to refer to a sequence of numbers arising from a word problem he copied into his book from another source. Meanwhile, for several centuries his true legacy hung on a single brief reference in one book (Pacioli's *Summa*).

Today, Leonardo's name is where it rightfully deserves to be, alongside the other great individuals who shaped our world and who we are. In the end, we do not need to travel to Italy to see Leonardo's legacy. We live it every day, every time we do something that depends upon the modern arithmetic he brought to the West.

Notes

Chapter 0: Your Days Are Numbered

1. Goetzmann 2004, p. i.
2. Cossali 1797, 1799.
3. Dauben and Scriba 2002, p. 69.

Chapter 1: A Bridge of Numbers

1. See Hughes 2008, p. 259.
2. The medieval mathematics scholar Rafaella Franci speculates that Bonacci may have been the name of Leonardo's grandfather, with Bonacci then meaning, in colloquial terms, the "father of the family". Personal communication, 2009.
3. See Butterworth 1999 or Devlin 2000.
4. Menninger 1992, p. 324.
5. Burnett 2005.
6. Ibid.
7. Burnett 2003.
8. An earlier suggestion that it was *Kitāb al-Jamʿ wa-l-tafrīq bi-ḥisāb al-Hind* (Book of addition and subtraction according to the Hindu calculation) is now viewed as incorrect, that being the title of a different book.
9. *Algoritmi*'s appearance in print was facilitated by Baldassarre Boncompagni, the same person who published the first printed edition of *Liber abbaci*.
10. There are five Arabic manuscripts extant, not all of them

complete. One complete Arabic copy is kept at Oxford, and a Latin translation is kept in Cambridge.

Chapter 2: A Child of Pisa

1. Much of the historical information in this chapter and the next is based on the pamphlet Gies and Gies 1969, now long since out of print.

Chapter 3: A Mathematical Journey

1. Only in the fifteenth century did a broader curriculum develop in Europe, with influences of the humanists.
2. Although the huge success of *Liber abbaci* in transforming the Western world was a result of its "do it this way" recipes for performing arithmetic and the vast range of examples it presented, many of them highly practical, it is nevertheless far more than a mere "arithmetic cookbook". Leonardo was a talented mathematician and a great believer in the formal, axiomatic approach to mathematics espoused by Euclid in his great work *Elements.* He took pains to develop the contents of *Liber abbaci* in a mathematically rigorous way, providing logical proofs to justify the methods he described.
3. Constable 2003, pp. 41, 221.
4. In particular, the contemporary historian of mathematics Barnabas Hughes commented on the matter in the introduction to his English-language translation of Leonardo's later book *De practica geometrie.* Hughes compared Leonardo's Latin quotations from *Elements* with all the extant Latin versions of the text available in the late eleventh century and found little agreement, leading him to suggest that the Pisan must have translated the quotations himself from an Arabic text (Hughes 2008, p. xix). Hughes concluded unequivocally, "Fibonacci was proficient in Arabic" (ibid.).

   There would likely have been no doubt about the matter were it not for a brief remark made by the scholar André Allard in 1984 that "Fibonacci's work does not show his knowledge of Arabic." (The English translation is in Allard 1996, p. 576.) Some scholars took that remark to mean that Leonardo did not know Arabic. But that is surely a misreading. Allard referred to "his knowledge", not "any knowledge", so the natural reading of the sentence is that Leonardo knew Arabic but his writing simply did not reflect that knowledge. (As well it should not; Leonardo wrote for a Latin-reading audience.) Besides, in a later work, Allard wrote, "Fibonacci used his knowledge of the Arabic Euclid to solve the problem posed above (*Elements*, II.I)" (Allard 2001, p. 88).

Chapter 4: Sources

1. Hughes 2008, p. xxii, suggests the following possibilities: *Kitāb al-Bayān wa at-tudhkār* (Book of demonstration and recollection) written by al-Hassār around 1175; the *Kitāb aal-kāfi fi 'īlm al-hisāb* (Sufficient book on the science of arithmetic) by al-Karajī; *al-Urjuza fīl-jabr wa-l muqābala* (Poem on algebra) by Ibn al-Yāsamin, which used verse as a means of recalling propositions and algorithms; Abū Kāmil's books *Kitāb fīl-jabr wa'l muqābala* (Book on algebra) and *Kitāb al-missāha wa'l-handasa* (Book of measurement and geometry), which did not survive to the present day. Also available were Arabic translations of Euclid's *Elements*, by al-Hajjāj ibn Matar (ca. 786–833), and by Ishaq al-'Ibādī (830–910), revised by Thabit ibn Qurra (836–901).
2. See Hughes 1986. In a private communication to me in 2010, Hughes claims there is evidence that Guglielmo also translated it into vernacular Italian.
3. Miura 1981.
4. Rosen 1831.
5. Ibid., p. xi.
6. All these quotations are taken from the al-Khwārizmī entry on the generally reliable online History of Mathematics archive run by the University of St. Andrews in Scotland: http://www-groups.dcs.st-andrews.ac.uk/~history/.
7. This is a common occurrence in the history of science, when an individual's expository work becomes famous and influential.

Carl Sagan, Richard Dawkins, and Steven Pinker are just three of many contemporary scientists whose popular expository writing came to overshadow their own original research.

8. In Rosen's translation there is one applied problem dealing with grain, but it was added to the text after al-Khwārizmī's time.
9. For more details on this, see Oaks and Alkhateeb 2007.
10. Shalhub 1990.
11. See page 48 of this book.
12. Oaks, personal communication, 2009.

Chapter 5: *Liber abbaci*

1. The translator, Laurence Sigler, uses the spelling "abaci" with one "b". I use Leonardo's spelling with two "b"s.
2. Sigler 2002, p. 148.
3. Ibid., p. 142.
4. Ibid., p. 220.
5. Ibid., p. 256.
6. Ibid., p. 317.
7. Ibid., p. 337.
8. Ibid., p. 337.
9. Ibid., p. 350.
10. Ibid., p. 362.
11. Ibid., p. 366.
12. Ibid., p. 128.
13. Ibid., p. 268.
14. Ibid., p. 460.
15. Ibid., p. 290.
16. Ibid., p. 291.

Chapter 6: Fame

1. Sigler 2002, p. 5.
2. An English translation was prepared by Barnabas Hughes and published by Springer-Verlag in 2008.

3. See Sigler 1987 for an English translation.
4. Lüneburg 1993.
5. Much of the historical detail in this chapter is taken from Gies and Gies 1969.
6. Sigler 1987, p.3.
7. Quadratic equations, of the form $ax^2+bx+c=0$, where the highest power of x is 2, are generally solved by a method known as "completing the square," typically taught in high schools today, but cubic equations, where there is an x-cubed, such as the one Leonardo was asked to solve, generally require a lot more work.
8. The translation is from Gies and Gies 1969, pp. 93–95.
9. See chapter 9.

Chapter 7: The Fibonacci Aftermath

1. Arrighi 1964, 1967, 1973, 1987.
2. Van Egmond 1980.
3. Heeffer 2008.
4. An English translation of the entire text is included in Swetz 1987.
5. Zervas 1975.
6. For a good general introduction to the abbacus schools, see Witt 2000.
7. See Van Egmond 1980, p. 8.
8. Eighteen faithful copies of *Trattato d'abacho* survive to this day.
9. Ulivi 2002, 2006.
10. Giovanni Villani, *Cronica*, vol. 6, pp. 184*f*, 1823. Some scholars have suggested that the actual figures were probably less than these, with the percentage of boys attending abbacus schools just 5 percent. See, for example, Grendler 1989, p. 72.
11. Swetz 1987, p. 182.
12. Struick 1968.
13. Swetz 1987, p. 182.
14. Those records are now housed in the Selfridge Collection at Harvard.
15. Struick 1987, pp. 80–81.
16. Ibid., pp. 182–83.

17. Swetz 1987, p. 6.
18. See Swetz 1987 for details. Just as it is no accident that Treviso saw the first printed textbook on practical arithmetic using the Hindu-Arabic system, so too when Franz Swetz published an English-language translation of the *Treviso Arithmetic* in 1987, accompanied by historical notes and a commentary, he pointedly titled his book *Capitalism and Arithmetic: The New Math of the 15th Century*.
19. Ibid., p. 297.

Chapter 8: Whose Revolution?

1. Høyrup 2005.
2. Van Egmond 1980, p. 31.
3. Gies and Gies 1969.
4. For instance, we know of two vernacular Italian translations of material in Leonardo's manuscript that appeared around 1350, one of chapters 14 and 15, the other of most of chapter 12 and a little of chapter 13. Another Italian translation of chapters 14 and 15 has been dated to around 1400.

   There were also vernacular versions of some of Leonardo's other works. An Italian translation of his *Liber quadratorum* also dates from 1400, and one of *De practica geometrie* is dated 1442.
5. This example and the summary of the abbacus books format I give in this chapter are taken from the introduction to Van Egmond 1980.
6. Franci 2003, p. 35.
7. Van Egmond 1980, p. 6.
8. Heefler 2007, p. 2.
9. Columbia X 511 A13. *Publications of the German Museum for the History of Science and Technology*, Series C, Source-texts and Translations, no. 33, Munich, 1977.
10. Travaini 2003.
11. Van Egmond 1980, p. 7.
12. Ibid., p. 31.
13. Vogel 1970, pp. 611–12.
14. Grattan-Guinness 2003.

15. Sigler 2002, p. 5.
16. Palatino 573, 433v, and 434r.
17. Ulivi 2002.
18. Page 154 of the Boncompagni edition of *Liber abbaci*.
19. In the Palatino 573 codex of the Biblioteca Nazionale of Florence (c.433v).
20. My description of this manuscript is based on Franci 2003, section 2.
21. Franci 2003, p. 35.
22. Antoni 1973, p. 334.
23. Franci 2003, p. 53.
24. It occupies sheets 1–70 of the Ricardian Codex 2252.

Chapter 9: Fibonacci's Legacy—in Stone, Parchment, and Rabbits

1. Sigler 2002, p. 404.
2. The translation is by Barnabas Hughes.
3. The details that follow are taken from an article by Rodolfo Bernardini, titled "Leonardo Fibonacci nella iconografia e nei marmi" (Leonardo Fibonacci in iconography and in marbles), published in the magazine *Pisa Economica*, 1977 (n.1), pp. 36–39. The article describes both the memorial stone and the statue. *Pisa Economica* is a magazine published by the Pisa Camera di Commercio, Industria, Artigianato (Chamber of Commerce).
4. Though some writers have assumed the park is named after Leonardo's friend Michael Scott, to whom he dedicated *Liber abbaci,* this is not the case. The name comes from Domenico Scotto, a Livornese shipowner who bought the property in 1798.
5. It is listed in the catalog as Conventi Sopressi C.1.2616.
6. It bears the reference mark Vatican Palatino #1343.
7. It has the reference number L.IV.20.

Bibliography

Ahmad, S., and R. Rashed, eds. *'Al-Bahir' en algèbre d'As-Samaw'al.* Damascus: University of Damascus Press, 1972.

Allard, A. "Arabic Mathematics in the Medieval West." In Roshdi Rashed, *Encyclopedia of the History of Arabic Science,* pp. 539–80. London: Routledge, 1996.

———. "The Influence of Abū Kāmil's Algebra on the Latin Authors of the 12th and 13th Centuries." *Journal for the History of Arabic Sciences* 12, nos. 1, 2 (2001): 83–90.

Anouba, A. "Samaw'al, Ibn Yaḥyā al-Maghribī al-." *Dictionary of Scientific Biography.* New York: Charles Scribner's Sons, 1970–90.

Antoni, T. "Le scuole d'abaco a Pisa nel secolo XIV." *Economia e Storia* 20 (1973): 334.

Arrighi, Gino, ed. *Antonio de' Mazzinghi. Trattato di Fioretti, secondo la lezione del codice L.IV.21 (sec. XV) della Biblioteca degli Intronati di Siena.* Pisa: Domus Galilaeana, 1967.

———, ed. *Libro d'abaco, Dal Codice 1754 (sec. XIV) della Biblioteca Statale di Lucca.* Lucca: Cassa di Risparmio di Lucca, 1973.

———, ed. *Paolo Dell'Abaco, Trattato d'aritmetica.* Pisa: Domus Galilaeana, 1964.

———, ed. *Paolo Gherardi, Opera mathematica: Libro di ragioni liber habaci. Codici Magliabechiani Classe XI, nn. 87 e 88 (sec. XIV) della Biblioteca Nazionale di Firenze.* Lucca: Pacini-Fazzi, 1987.

Ashtor, Eliyahu. *Levant Trade in the Later Middle Ages.* Princeton: Princeton University Press, 1983.

Bernardini, Rodolfo. "Leonardo Fibonacci nella iconografia e nei marmi" (Leonardo Fibonacci in iconography and in marbles). *Pisa Economica* 1977 (n.1), Pisa Camera di Commercio, Industria, Artigianato (Chamber of Commerce), pp. 36–39.

Boncompagni, Baldassarre. *Scritti di Leonardo Pisano*, Vol. 1 (*Liber abbaci*), 1857; vol. 2, 1862.

Burnett, Charles. "Leonardo of Pisa (Fibonacci) and Arabic Arithmetic." MuslimHeritage.com, 2005, http://www.muslimheritage.com/topics/default.cfm?articleid=472.

———. "The Transmission of Arabic Astronomy via Antioch and Pisa." In Jan P. Hogendijk and Abdelhamid I Sabra, eds., *The Enterprise of Science in Islam: New Perspectives*, pp. 36–37. Cambridge, Mass.: MIT Press, 2003.

Butterworth, Brian. *What Counts: How Every Brain Is Hardwired for Math*. London: Free Press, 1999.

Constable, Olivia. *Housing the Stranger in the Mediterranean World: Lodging, Trade, and Travel in Latin Antiquity and the Middle Ages*. Cambridge: Cambridge University Press, 2003.

Cossali, Pietro. *Origine, trasporto in Italia, primi progressi in essa dell'algebra. Storia critica di nuove disquisizioni analitiche e metafisiche*. Parma, 1797, 1799.

Dauben, Joseph, and Christopher Scriba. *Writing the History of Mathematics: Its Historical Development*. Basel: Birkhäuser, 2002.

Devlin, Keith. *The Math Gene: How Mathematical Thinking Evolved and Why Numbers Are Like Gossip*. New York: Basic Books, 2000.

Favier, Jean. *Finance and fiscalité au Bas Moyen Âge*. Paris: Société D'Édition D'Enseignement Supérieur, 1971.

Fisher, Irving. *The Theory of Interest*, New York: Macmillan, 1930.

Franci, Rafaella. "Leonardo Pisano e la trattatistica dell'abaco in Italia nei secoli XIV e XV." *Bollettino di Storia delle Scienze Mathematiche*, Intituti Editoriali e Poligrafici Internazionali, vol. 23, 2003, facs. 2, pp. 33–54.

Gandz, S. *The Mishnat Ha Middot: The First Hebrew Geometry of About 150 C. E. and; The Geometry of Muhammad Ibn Musa Al-Khowarizmi: the First Arabic Geometry (C. 820), Representing the Arabic Version of the Mishnat Ha Middot*. Berlin: Julius Springer, 1932.

Gies, J., and F. Gies. *Leonardo of Pisa and the New Mathematics of the Middle Ages*. New York: Crowell Press, 1969.

Goetzmann, William N. *Fibonacci and the Financial Revolution.* National Bureau of Economic Research, Working Paper 10352, March 2004.

———. *Fibonacci and the Financial Revolution.* In Goetzmann and Rouwenhorst 2005, pp. 123–43. (A revised version of Goetzmann 2004.)

Goetzmann William N., and K. Geert Rouwenhorst, eds. *Origins of Value: The Financial Innovations that Created Modern Capital Markets.* New York: Oxford University Press, 2005.

Graham, John R., and Harvey Campbell. "The Theory and Practice of Corporate Finance: Evidence from the Field." *Journal of Financial Economics* 60, nos. 2–3 (2001): 187–243.

Grattan-Guinness, Ivor. *Companion Encyclopedia of the History and Philosophy of the Mathematical Sciences.* Vol. 1, p. 201. London: Routledge, 1992.

Grendler, Paul. *Schooling in Renaissance Italy: Literacy and Learning, 1300–1600. The Johns Hopkins University Studies in Historical and Political Science,* 1989, p. 72.

Grimm, Richard E. "The Autobiography of Leonardo Pisano." *Fibonacci Quarterly* 11, no. 1 (February 1973): 99–104.

Heefler, Albrecht. "The Abbacus Tradition: The Missing Link between Arabic and Early Symbolic Algebra." *Proceedings of the International Seminar on the History of Mathematics.* New Dehli, India, 2007.

———. *The Abbaco Tradition (1300–1500): Its Role in the Development of European Algebra.* ScientificCommons.org, 2009.

Herlihy, David. *Pisa in the Early Renaissance: A Study of Urban Growth.* New Haven: Yale University Press, 1958.

Hogendijk, Jan P., and Abdelhamid I Sabra, eds. *The Enterprise of Science in Islam: New Perspectives,* pp. 36–37. Cambridge, Mass.: MIT Press, 2003.

Høyrup, Jens. "Jacopo da Firenze and the Beginning of Italian Vernacular Algebra." International Workshop on *The Origins of Algebra: From al-Khwārizmī to Descartes,* Barcelona, March 27–29, 2003. Roskilde University, Section for Philosophy and Science Studies, preprint no. 6, 2003. Published in *Historia Mathematica 33* (2006): 4–42.

———. *Jacopo da Firenze's* Tractatus Algorismi *and Early Italian Abbacus Culture.* Basel: Birkhäuser, 2007.

———. "Leonardo Fibonacci and *abbaco* Culture: A Proposal to Invert the Roles." *Revue d'Histoire des Mathématiques* 11 (2005): 23–56.

Hughes, Barnabas. "Gerard of Cremona's Translation of al-Khwarismi's *Al-jabr:* A Critical Edition." *Mediaeval Studies* 48 (1986): 233.

———. "Fibonacci, Teacher of Algebra: An Analysis of Chapter 15.3 of Liber Abbaci." *Mediaeval Studies* 64 (2004): 313–61.

———. *Fibonacci's* De practica geometrie. New York: Springer-Verlag, 2008.

Khalil, Roshdi, trans. *Algebra wa al-muqabala,* by Omar Khayyam. United Kingdom: Garnet, 2008.

Livio, Mario. *The Golden Ratio.* New York: Broadway Books, 2002.

Lüneburg, Heinz. *Leonardo Pisani* Liber Abbaci *oder Lesevergnügen eines Mathematikers.* Mannheim, Germany: B. I. Wissenschaftsverlag, 1993.

Markowsky, George. "Misconceptions About the Golden Ratio." *College Mathematics Journal* (January 1992): 2–19.

Menninger, Karl. *Number Words and Number Symbols: A Cultural History of Numbers.* New York: Dover, 1992.

Miura, Nobuo. "The Algebra in the *Liber abaci* of Leonardo Pisano." *Historia Scientiarum* 21 (1981): 57–65.

Munro, John H. "The Medieval Origins of the Financial Revolution: Usury, Rentes, and Negotiability." *International History Review* 25, no. 3 (September 2003): 505–62.

Oaks, Jeffrey A., and Haitham M. Alkhateeb. "Simplifying Equations in Arabic Algebra." *Historia Mathematica* 34, issue 1 (February 2007): 45–61.

Parshall, K. H. "The Art of Algebra from al-Khwarizmi to Viète: A Study in the Natural Selection of Ideas." *History of Science* 26, no. 72 (1988): 129–64.

Pezzolo, Luciano. "Bonds and Government Debt in Italian City States: 1250–1650." In Goetzmann and Rouwenhorst, eds., *Origins of Value: The Financial Innovations that Created Modern Capital Markets.* New York: Oxford University Press, 2005.

Pryor, John H. "The Origins of the Commenda Contract." *Speculum* 52, no. 1 (January 1977): 5–37.

Rashed, Roshdi. *The Development of Arabic Mathematics: Between Arithmetic and Algebra*. London: Kluwer, 1994.

———. "Fibonacci et les mathématiques arabes." *Micrologus* 2 (1994b): 145–60.

Rosen, Frederic. *The Algebra of Mohammed Ben Musa*. London: J. Murray, 1831.

Shalhub, S. "The Calculations and Algebra of Abū Kāmil Shuja ibn Aslam and His Effects on the Work of al-Karaji and on the Work of Leonardo Fibonacci" (Arabic). *Deuxième Colloque Maghrebin sur l'Histoire des Mathématiques Arabes* (Tunis, 1990), A23–A39.

Sigler, Laurence. *The Book of Squares: An Annotated Translation into Modern English*. New York: Academic Press, 1987.

———. *Fibonacci's Liber Abaci: A Translation into Modern English of Leonardo Pisano's Book of Calculation*. New York: Springer Verlag, 2002.

Sivéry, Gérard. "Mouvements de capitaux et taux d'interêt en occident au XIIIe siècle." *Annales Economies Societés Civilizations* 38 (1983): 367.

Smith, D. E. *History of Mathematics*. Vol. 1. New York: Dover, 1951.

Spufford, Peter. *Power and Profit: The Merchant in Medieval Europe*. New York: Thames and Hudson, 2002.

Struick, Dirk J. *A Concise History of Mathematics*. 4th ed. New York: Dover, 1987.

———. "The Prohibition of the Use of the Arabic Numerals in Florence." *Archives Internationales d'Histoire des Sciences 21* (1968): 291–94.

Swetz, Frank J. *Capitalism and Arithmetic: The New Math of the 15th Century*. LaSalle: Open Court, 1987.

Toomer, G. J. Biography in *Dictionary of Scientific Biography*. New York: Scribner, 1970–90.

Travaini, Lucia. *Monete, mercanti e matematica, le monete medievali nei trattati di aritmetica e nei libri di mercatura*. Rome: Jouvence, 2003.

Ulivi, Elisabetta. "Benedetto da Firenze (1429–1479), un maestro d'abbaco del XV secolo. Con documenti inediti e con un' appendice su abacisti e scuole d'abaco a Firenze nei secoli

XIII–XVI." *Bollettino di Storia delle Scienze Matematiche* 22 (2002): 3–243.

———. "Raffaello Canacci, Piermaria Bonini e gli abacisti della famiglia Grassini." *Bollettino di Storia delle Scienze Matematiche* A. 24, n. 2 (2004), stampa 2006, pp. 125–212.

Van Egmond, Warren. *Practical Mathematics in the Italian Renaissance: A Catalog of Italian Abbacus Manuscripts and Printed Books to 1600.* Florence: Instituto E Museo di Storia Della Scienza, 1980.

Viète, François. *In artem analyticam isagoge.* Tours, 1591. Reprinted in Viète, *Opera Mathematica.* Leiden, 1646; London, 1970.

Vogel, Kurt. "Fibonacci, Leonardo or Leonardo of Pisa." *Dictionary of Scientific Bibliography.* Vol. 4, pp. 603–13. New York: Scribner, 1970.

Witt, Ronald G. *In the Footsteps of the Ancients: The Origins of Humanism from Lovato to Bruni.* Leiden: Brill, 2000.

Zervas, Diana Finiello. "The *Trattato dell'Abbaco* and Andrea Pisano's Design for the Florentine Baptistry Door." *Renaissance Quarterly* 28 (1975): 483–503.

Index

Note: page references followed by n refer to notes, with note number where appropriate.

A Note on the Author

KEITH DEVLIN is a senior researcher and executive director at Stanford's H-STAR Institute, which he cofounded. He is also a cofounder of the Stanford Media X research network. NPR's "Math Guy," he is the author of thirty books, including *The Math Gene*. He lives in Palo Alto, California.